Texas Water Recreation

Don't miss these other Roadrunner Guides:

Texas Zoos and Animal Parks
Texas Flea Markets *(available Spring 1991)*
Great Texas Getaways *(available Spring 1991)*

Texas Water Recreation

A Roadrunner Guide

by
Ann Ruff

Taylor Publishing Company
Dallas, Texas

Book designed by Deborah J. Jackson-Jones
Series logo designed by Morren Design

Copyright © 1990 by Ann Ruff

All rights reserved.

No part of this book may be reproduced in any form without written permission from the publisher.

Published by Taylor Publishing Company
               1550 West Mockingbird Lane
               Dallas, Texas 75235

Library of Congress Cataloging-in-Publication Data

Ruff, Ann, 1930-
    Texas water recreation/by Ann Ruff.
       p. cm.—(A Roadrunner guide)
    ISBN 0-87833-627-3: $9.95
    1. Outdoor recreation—Texas—Guide-books. 2. Aquatic sports—Texas—Guide-books. 3. Recreational areas—Texas—Guide-books. 4. Texas—Description and travel—1981—Guide-books. I. Title. II. Series.
GV191.42.T4R83 1990
333.78'4'09764—dc20   89-49274
                                                                      CIP

Printed in the United States of America

10 9 8 7 6 5 4 3 2 1

To Richard Wright for dragging my canoe, swimming in Barton Springs, and braving The Hydra.

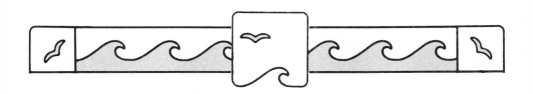

CONTENTS

Introduction   1

Courtesy of Mother Nature: Texas Swimming Holes   3

    Balmorhea   5
    Barton Springs   7
    Blanco State Recreation Area   9
    Blue Hole   11
    Blue Lagoon   13
    Boykin Springs Recreation Area   16
    Crystal Springs Beach   18
    Dinosaur Valley State Park   20
    Fort Clark   22
    Hamilton Pool   25
    Krause Springs   27
    Lake Tejas   29
    Little Arkansas (Lil Ark)   31

Making a Splash: Texas Water Parks   33

    Fame City Waterworks   35
    Red Cloud Water Park   37
    Schlitterbahn   39
    Splashtown   42
    Splashtown USA   45
    Summerfun USA   48
    WaterWorld   50
    Wet 'N Wild   53

A Beach Within Reach: Texas Beaches   57

     Galveston   61
     North Padre Island   68
     Rockport—Port Aransas   71
     South Padre Island   77
     Surfside Beach   82

How to Be a Texas River Rat: Texas Rivers   85

     Big Thicket   89
     Brazos River   93
     Comal River   96
     Guadalupe River   100
     Guadalupe River State Park   105
     Pedernales Falls State Park   107
     Rio Frio   109
     Rio Grande   112
     Sabinal River   116
     San Marcos River   118
     Upper Guadalupe River   123

Lakes Superior: Texas Lakes   127

     Lake Amistad   129
     Caddo Lake   132
     Crystal Lake   136
     Highland Lakes   138
     Lake Livingston   144
     Lake Texoma   147

KEY TO SYMBOLS

 Camping

 RV Hookup

 Picnic Area

 Concessions

 Showers

 Rest Rooms

 Lifeguard

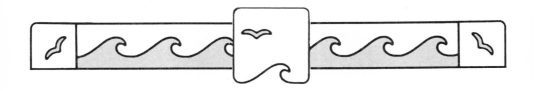

INTRODUCTION

Texas Water Recreation is one of the first of a new series of travel guides, the Roadrunner Guides, that will cover the Lone Star State. Containing the latest detailed information for both the native Texan and the out-of-state visitor, these handy guides offer in-depth looks at subjects dear to almost every traveler's heart, subjects seldom covered adequately by more traditional travel texts.

Texas Water Recreation shows you how to beat the heat at the many water recreation sites throughout the state. From Gulf Coast beaches to out-of-the-way swimming holes, you'll find everything you need to know about cooling off during the long Texas summer.

So—come on in, the water's fine!

Courtesy of Mother Nature: Texas Swimming Holes

Last one in is . . . who wants to be a rotten egg? Read on and discover the best swimming holes in Texas.

When you were a kid, maybe yours was an abandoned gravel pit, or a deep bend in the nearby river, or even a tank on a ranch—all highly unsanitary, of course. But somehow you never caught anything except a few chiggers, and your eyes never turned red or your hair green from too much chlorine, either. Today the old swimming holes are just about gone. Tract houses have moved in, water pollution has seeped up, and many people have chosen the status symbol of a kidney-shaped pool in the back yard.

If you grew up in a small town or on a farm, you knew the fun of hiking out to the gravel pit with sweat streaming from every pore in your body. And finally you were there! At last! Cool water! If you were by yourself or with the gang of just your own sex, clothes were shucked off to cries of, "Last one in is a rotten egg." Nobody wanted to be the rotten egg. Skinny-dipping was the best of all worlds. Maybe there was an exciting rope swing out over the water, and the trick was to see who could swing out the farthest. Diving for rocks or pennies wasn't too great because the muddy water made them too hard to find, but there was lots of splashing and playing around.

After stretching out the day as long as possible, you dried off in the sun. Nobody ever brought a towel. With hair still damp and shoes filled with sand, it was time to go home.

Sometimes groups of boys and girls would get together a gravel-pit picnic with no skinny-dipping allowed. Girls brought the food, and the guys brought the drinks. Somebody always had a truck to use for a hayride out to the swimming hole, and somebody always knew how to build a fire for the hot dogs and marshmallows. On these occasions, everybody brought towels and combs, too. As soon as the sun went down, the fire was lit, and hot dogs were stuck on sticks. The locusts began to sing, the fireflies put on their little light show, and the mosquitoes prepared for their feast, too. No one ever remembered the bug repellent.

If you grew up without a swimming hole, you missed a wonderful part of adolescence. A healthy clean chlorinated concrete pool just doesn't do a thing for nostalgia. It's those memories of your own private place that linger forever. A few of the good ol' swimming holes are still to be found around Texas complete with rope swings, muddy bottoms, and rocky banks; but, alas, no skinny-dipping.

BALMORHEA

WEST

Balmorhea State Recreation Area
Box 15
Toyahvale 79786
915-375-2370

Location: Balmorhea is just off I-10 47 miles west of Fort Stockton

Admission Fee

Hours: Open 7 days 8:00 a.m.–10:00 p.m. from the fourth Friday in May through Labor Day

Amenities: Rest rooms, bathhouse, concessions, lifeguard, cabins, campsites, trailer sites

Have you ever wanted to go swimming in the largest manmade swimming pool in the United States? Well, Balmorhea covers one and three-quarters acres and holds more than 3.5 million gallons of water from San Solomon Springs. Okay, so it's big. The water is a pleasant 76 degrees, the setting is absolutely lovely, and you have all the ingredients for a superstar swimming hole.

Balmorhea was named by a firm of land promoters, Balcolm, Morrow, and Rhea, who combined their names to create the town. Incidentally, it's pronounced bal-mo-ray. There's not much town anymore, but thankfully the springs endure and the pool is fed at the rate of 26 million gallons of water per day.

Of course the Mescalero Apaches knew about the springs, and they were once a source of water for Mexican farmers. Then in the thirties the Civilian Conservation Corps arrived and did a very good job of modernizing the area, making it an ideal recreational site. A Spanish-style bathhouse was

Courtesy of Mother Nature

built, banks were cemented, and the grounds were landscaped. The gravel bottom was left in its natural state, and the color of the water is almost dazzling in the hot sun. Amazingly, the color is comparable to the brilliant turquoise of the Caribbean.

This circular pool is so large you can hit the high board and swim head-on for 750 feet to the pavilion at the edge of the springs. Not far away you can see the Davis Mountains, a perfect backdrop for Balmorhea. Naturally, all of this water and scenery is a magnet to those poor souls baking in the heat of the Trans-Pecos and Permian Basin. Some Sundays more than a thousand people come to be refreshed, but there is still plenty of room. That's how huge the pool is.

You share your swim with two resident endangered species, the Comanche Springs pupfish and the Pecos mosquito fish. Don't worry. They are very friendly and are glad you are here.

It's hard to believe, but there are actually a few complaints that the water is not chlorinated. But out in this desert, where would the pollution come from? Anyway, chemicals will never ruin Balmorhea because after the water leaves the pool it is used for irrigation. The water in the pool is changed every six hours by the tremendous flow of the springs.

Plan to spend a night (you no doubt came a long way) at the San Solomon Springs Courts, which are part of the recreation area. The cabins have central heat and air conditioning, kitchenettes, and even TV. Of course, it's much more romantic to sit out and watch the stars—and in this clear air, how bright they shine! McDonald Observatory is down the road in Fort Davis because of these clear skies. During the summer, reservations are a must at Balmorhea.

BARTON SPRINGS
Zilker Park
Austin 78704
512-476-9044

HILL COUNTRY

Location: Barton Springs Road between Robert E. Lee and Loop 1

Admission Fee

Hours: 6:00 a.m.–10:00 p.m. daily

Amenities: Rest rooms, bathhouse, lifeguards, free parking

Actually, Austin's famous springs should be called Parthenia and Eliza Springs, because that's what Billy Barton named them. Uncle Billy was sort of the Daniel Boone of Texas and lived for a while near La Grange. He complained that his neighbors ten miles away were too close, so before 1837 Barton moved and built a home near the springs that were to be named for him. Somehow Parthenia and Eliza, his two daughters, lost out.

Indians and Spanish explorers had liked the springs, too, even though Uncle Billy ended up with them, and in 1917 the city of Austin purchased the springs for a municipal park. You don't have to worry about the water being polluted during dry periods, since the flow varies from a minimum of fifteen million to as much as forty-two million gallons daily, but unfortunately an inch or more of rain will cause an urban runoff that is hazardous to your health.

In the 1930s Barton Creek was dammed and the concrete walls added, creating a pool 950 feet long. All 950 feet maintain a steady 68- (br-r-r-r) degree temperature even when Austin is sizzling and the humidity soaring. You can always count on the icy water at Barton Springs to cool you

Courtesy of Mother Nature

off. In fact, on a hot weekend three to four thousand people come through the gate, so why not avoid the rush and go early in the morning or late in the afternoon? Or you can join those extremely hearty souls who swim here all winter and really have the springs to yourself. Other spring-fed pools claim a 68-degree water temperature, but Barton Springs seems to be the coldest of them all. No wonder the hillsides have more people on the grassy slopes than there are in the water. You don't even have to go in the water to cool off—just looking at it is enough.

Barton Springs is an Austin tradition. Even if you just loll in the shade of the old pecan trees instead of letting that cold water shock your senses, it is *the* place to be during an Austin summer.

BLANCO STATE RECREATION AREA

HILL COUNTRY

Box 493
Blanco 78606
512-833-4333

Location: 1 mile south of Blanco on U.S. 281

Admission Fee

Amenities: Camping, RV hookups, rest rooms, picnic area, paddleboat and canoe rentals

Restrictions: No lifeguards

As with all of the beautiful Hill Country rivers, the Blanco rises from limestone seeps and springs. Christened Blanco (White) in 1721 by a Spanish explorer, the Marquis de Aguayo, the river was considered as a site for a mission, but a location farther north on the San Saba River was chosen. The Blanco is a shallow, swift stream that cuts a spectacular route near the Devil's Backbone portion of the Hill Country.

The rock ledges and outcrops belong to the Glen Rose formation of Cretaceous age, which is noted for its dinosaur tracks (see Dinosaur State Park). The big ugly critters didn't leave their footprints behind here at the park, but if you want a three-mile trek down the river, there they are. Ask the park superintendent for information and advice before you go.

The Blanco River has been dammed here, and the result is a nice swimming hole. You wouldn't choose it to train for the racing team, but for a lot of cool splashing, it is a treat. Try above the dam, just west of the park entrance. Even though this is a spring-fed river, the visibility in the dark green water is limited to just a few feet.

Courtesy of Mother Nature

This part of the river isn't overwhelmingly scenic, but it's one of those spots in the Hill Country about which when you are passing by, you think, "Why don't I just stop and have a Hill Country bracer?"

Paddleboats and canoes are for rent, or you can take it easy on the manicured grounds that look like someone's back yard. The sloping banks are shaded with willows and oaks, and barbecue pits are plentiful. Drag out your lawn chair and watch the traffic on U.S. 281 whiz by.

The little town of Blanco just to the north of the park is sort of quaint. Its historic courthouse squats dejectedly on an acre or two right on the highway. Abandoned and aging badly, it's for sale. If you've always wanted your own courthouse, there is a bargain in Blanco.

BLUE HOLE

P.O. Box 331
Wimberley 78676
512-847-9127

HILL COUNTRY

Location: A quarter-mile east of Wimberley on Ranch Road 3237

Admission Fee (memberships required)

Amenities: Rest rooms, camping, RV hookups

Restrictions: No lifeguards

How many Blue Holes do you suppose there are in Texas? They may not be very blue, but you can bet that wherever there's a river, kids have discovered their favorite "Blue Hole." Well, Blue Hole at Wimberley is actually a deep blue, and your skin will be the same color after a few minutes in the 65-degree water. Talk about a refreshing dip!

Wimberley's Blue Hole has been popular for years. During World War II, troops from San Antonio used it for R and R, and it was a favorite spot for cooling off as far back as the twenties. Needless to say, this was before Wimberley became the artsy-craftsy town it is today. Since tourists now love Wimberley, head for Blue Hole early.

Blue Hole is actually a narrow stretch of Cypress Creek covered by (guess what) cypress trees. Very little sun cuts through the leaves to warm up the water. Two wonderful rope swings can easily handle all the excess energy the kids have, and tree ladders for diving can work off stored-up calories. As for actually putting in your quota of laps, the swimming section is only about 250 feet long. How about just forgetting your workout and lying on a float with a cool drink? That's really what Blue Hole is all about. If the water

Courtesy of Mother Nature

is too cold for you, bring your lawn chairs and enjoy the lovely setting.

As for the membership fee (anyone can buy a membership), the owners of Blue Hole say, "In a move to protect the area and avoid insurance problems that are driving others out of business, Blue Hole has become a recreation club. The change has been brought on by tax legislation and the rising cost of insurance. In addition, years of heavy usage has begun to show its wear and tear on the area."

Memberships are very reasonable, and they haven't lessened Blue Hole's popularity one whit.

BLUE LAGOON

16405 Wall Street
Houston 77040
Richard Grifno 713-376-3157
Huntsville 409-291-6111

SOUTHEAST

Location: Take Exit 123 off of I-45 just 6 miles north of Huntsville. Go east 4 miles, and watch for flags.

Admission Fee

Hours: Weekends 8:00 a.m.–dark
Weekdays May–Oct.; 10:00 a.m.–6:00 p.m.

Amenities: Rest rooms, concessions, camping, air-fill station, picnic tables, shelters for divers (reservations requested), first-aid station (including oxygen)

Restrictions: No swimming, no pets, no children under twelve, no glass containers; cliff-jumping will result in expulsion

It took millions of years for *Homo sapiens* to come out of the seas and walk on land. Now we spend millions of dollars trying to return to the sea. Since our gills have disappeared we now replace them with an air tank; plastic flippers have become fins; and scales and blubber have turned into wet suits. It all makes scuba diving one of the most popular sports in Texas.

There isn't a scuba diver worth his or her flippers who hasn't been to the incredibly clear waters off the island of Cozumel near the Yucatan Peninsula. Yet, there is a tiny bit of Cozumel at the Blue Lagoon in the Piney Woods of East Texas.

Courtesy of Mother Nature

It's not easy to think of a lagoon in East Texas. But if you go back thirty million years or so, the seas did reach this far into Texas, and countless fossils were left behind. The remains of the prehistoric plant and animal life formed the quartzite which was mined for the Galveston jetties. Hundreds of artesian springs were discovered that created a perfect environment for open-water scuba training.

Actually, Blue Lagoon is two lagoons (each about six acres) of clear turquoise water with forty-foot visibility. Large platforms twenty-feet square are located in each lagoon at approximately twenty-foot depths to provide ample space to practice safe diving skills. In addition to the spectacularly beautiful underwater quarries, divers can access submerged yachts placed here to be explored. No, they aren't Spanish galleons, and you won't discover any sunken treasure, but you'll have a lot of fun.

In addition to student classes, several dive clubs use the lagoons for their events. The submerged wrecks become haunted boats at Halloween, and about 350 divers will search for hidden pumpkins and shake hands with the skeleton sitting on the toilet. Then the Easter Bunny hides his eggs around the wrecks. Handicapped divers really enjoy Blue Lagoon, and the Harris County Sheriff's Department holds its dive classes here.

The secret of the lagoon's clear water is its low pH, too low to support the growth of algae or fish. That means you can't watch the exotic tropical fish that inhabit the bright waters off of Cozumel, or even spy a fat Texas catfish, but you won't get infected ears either. Underwater photographers love Blue Lagoon, as the clear water makes it a lot easier to test new techniques or equipment. Also, it's a great tune-up spot for divers who are planning a big international trip and haven't had their gear in the water for a while.

You won't be doing much diving at the lagoons in the winter, because that pretty water drops to a brisk 47 degrees in January. But in February the temperature starts building, and reaches 92 degrees in August. Don't worry; the springs more than offset the evaporation factor.

For dive clubs, dive groups, and just plain old certified divers, holiday weekends are good times to come and enjoy these unique lagoons. These are the days they do the least business with open-water training, so there is plenty of room to get your gear and body in tiptop diving condition.

Keep these special diving spots as safe as possible by following some simple rules. Leave Spot and Rover at home. Kids have to be over twelve years old to get in a scuba tank. Absolutely no swimmers are allowed. These lagoons are for divers only. If you are thinking of cruising up on your motorcycle or bringing your three-wheeler, forget it. Underwater scooters are allowed, but only on weekdays.

If you want to "rough it" the easy way, Huntsville State Park is nearby with tent and RV campsites and overnight cabins (409-295-5644). If a comfortable motel is your style, the big chains are in Huntsville. You might ask at Blue Lagoon if any are giving divers special rates.

While Blue Lagoon is often called "The Texas Cozumel," you probably won't be happy until you see the Mexican Cozumel. Yet, Blue Lagoon is especially suited for that first open-water experience, as well as a wonderful opportunity for continuing every diver's further education in the sport.

As one of the divers said, "Best of all, the place is run by divers for divers, so it's very conducive to our accomplishing whatever we need to do."

Courtesy of Mother Nature

BOYKIN SPRINGS RECREATION AREA
Angelina National Forest

EAST

Location: 13 miles southeast of Zavalla off Texas 63

Admission Fee

Amenities: Rest rooms, showers, camping, picnic area

Restrictions: No lifeguards

When the Good Lord created East Texas, He said, "I'm going to make an especially pretty spot where people can really appreciate My talents. I'm stocking it with some of My most beautiful birds, tallest trees, and loveliest plants. Many wild animals will make it their home, and I will make the forests so deep that the animals can hide from the people. Its name will be Angelina National Forest and, hopefully, man will cherish its beauty forever."

Unfortunately, man did not follow God's wishes, and much of the virgin timber was cut or knocked down by the skidders—they skidded the big trees out—before the 1930s. But during the Great Depression, the Civilian Conservation Corps came in, and thanks to the Corps, the forest is again the way God planned.

In addition to an "ooh" and "aah" dip in Boykin Springs, you can hike through the woods and discover an old ghost town where a sawmill once thrived, picnic under a gorgeous canopy of trees, or cast your line in the far end of the little lake. Maybe a fish or two will be curious enough to nibble.

A few years ago a rope swing hung from the limb of a tall pine, and you can imagine what fun that was. But sadly, rope swings have fallen victims to high insurance rates, and it is the rare swimming hole where kids can still pretend they are Tarzan. Even without its rope swing, Boykin Springs has the charisma of the ol' swimming hole.

As for a bit of history, Boykin Springs took its name from a pioneer named Sterling Boykin who settled here in the 1840s. The family cemetery is located near the springs, and Sterling rests amid his beloved forests. Just before he died, the old settler gave the land to the government. Thanks to God and Sterling Boykin we can enjoy nature at its absolute best.

Courtesy of Mother Nature

CRYSTAL SPRINGS BEACH

NORTHEAST

Box 390
Maud 75567
214-585-5246

Location: Three miles west of Maud on U.S. 67

Admission Fee

Hours: 10:00 a.m.–8:00 p.m. daily
Memorial Day through Labor Day

Amenities: Lifeguards, concessions, camping, pavilions, rest rooms

Shelly and Leon Jennings, owners of Crystal Springs Beach, boast that it is "as close to a Florida beach as you can get without actually being there!" Well, it's kind of hard to imagine a Florida beach surrounded by an East Texas forest filled with opossums, raccoons, rabbits, and even a lonesome coyote. But the sand is white, even if it was hauled in from nearby Old Boston, Texas. You may be a long way from Florida, but you will have a great time at the clear twelve-acre lake at Crystal Springs Beach.

Ever wanted to take a flying carpet ride? Well, you can pretend you're Aladdin and hop aboard. Your trip takes you 400 feet from start to finish, including one full turn and two half turns. This unique carpet is propelled by 1,500 gallons of water per minute. You don't even have to rub a magic lantern for this big thrill, but you have to climb up, up, up.

After your carpet deposits you safely, how about the Thrill Cable or the newest attraction, the Cannonball Ride? The cannonball shoots you through tubes, while maybe not at the speed of a silver bullet, you go *fast!* At the Thrill Cable, you launch yourself from a 30-foot tower and glide down a

300-foot cable to drop into deeper water. The Thrill Cable is an old attraction, but it's always popular.

As you swim out to the large diving platforms, don't be surprised if you feel a nip or two, because small creek perch may think you are a bonanza lunch treat. Just ignore them, and they will realize you are definitely inedible. Or why not rent a paddleboat and give your legs a good workout as you cruise the lake?

For youngsters, there is a shallow water area and a very attentive lifeguard, so you can relax and watch them love the water, too.

Crystal Springs Beach first opened for business in 1947 when cokes were a nickel and ice cream a dime. On its first Fourth of July more than a thousand water lovers swam and picnicked, and even though it has seen several owners, this has always been a wonderfully family-oriented vacation spot. So come on in, the water is fine!

Courtesy of Mother Nature

DINOSAUR VALLEY STATE PARK

NORTH CENTRAL

P.O. Box 396
Glen Rose 76043
817-897-4588

Location: 3 miles west of Glen Rose on U.S. 67

Admission Fee

Hours: Dinosaur Exhibit 8:00 a.m.–5:00 p.m. daily

Amenities: Camping, rest rooms

Restrictions: No lifeguard

Have you ever wanted to swim where dinosaurs roamed? At Dinosaur Valley, you can not only swim but walk in their footsteps. The Paluxy River is little more than a trickle in the summer months and there, frozen in limestone, are birdbath-size tracks of the "terrible lizards" who once found Texas a great place to live. Of course, that was about 105 million (give or take a millon) years ago.

Unlike bones, footprints are irrefutable proof of an animal's activities in a specific locality. In some mysterious way, scientists know that *Acrocanthosaurus* was traveling at about 5 miles per hour as he pursued the more ponderous *Pleurocoelus* who could only lumber along at 2.7 miles per hour. Scientists also suppose that *Pleurocoelus* ended up as dinner for *Acrocanthosaurus*. The third category of prints found at Dinosaur Valley is a mystery guest, but his identity will no doubt be discovered as more and more information is found on these big critters.

There used to be lots of dinosaur prints in the Paluxy's bed, but many were quarried and carried away. The Texas

Memorial Museum in Austin has a big slice of the prints, as does the Smithsonian. When these prints were discovered back in 1909, nobody cared, and big chunks of limestone were sold by the side of the road. Now, they are highly prized by scientists, and great care is taken to see that the Paluxy River bed is not destroyed.

Not far from the park entrance is a life-size *Brontosaurus* and a *Tyrannosaurus rex*—fiberglass, of course. You'll definitely want to line up the family and pose for photographs with the big ugly guys. Even though these two Gargantuas of the Jurassic Age didn't like the Texas scene and actually lived elsewhere, they add to the park's whole dinosaur ambience, as do the exhibits at the visitor center.

Now, where to swim? The Paluxy makes a deep bend in the park that is the Blue Hole of the river. The water is clean and cool even in the summer, and while it's not very deep, you can do enough strokes to relax from your lessons in prehistoric history. It is kind of neat to think that your footprints may become fossils, and that millions of years from now creatures from outer space may puzzle over how fast you were running to get to camp for dinner.

If you are not a camper, stay at Glen Rose's excellent Inn on the River (817-897-2101). Also, don't miss Fossil Rim Ranch. You can drive through this exotic animal ranch and see rare animals which are well on their way to extinction unless they are saved by man.

Courtesy of Mother Nature

FORT CLARK
P.O. Box 1401
Brackettville 78832
512-563-2494

RIO GRANDE VALLEY

Location: 123 miles west of San Antonio on U.S. 90

Admission: free to guests

Hours: 10:00 a.m.–6:00 p.m.

Amenities: Rest rooms, lifeguards from Memorial Day to Labor Day, picnic area, motel, RV park, restaurant, golf courses

Restrictions: Resort guests only allowed

The life span of most Texas frontier forts was very brief, but Fort Clark survived long after the threat of Indian warfare was over. Activated in 1852, it was June of 1944 before the closing of the fort was announced. It was at Fort Clark that the famous Seminole-Negro Indian Scouts were stationed, and the fort was the base of operations for Colonel Ranald S. Mackenzie's raids into Mexico to punish renegade Indians. Four of the Seminole-Negro Indian Scouts serving under Colonel John Bullis received the Medal of Honor for their bravery against the Comanches and Apaches.

Names famous in American history were stationed at Fort Clark as well. Among them were General George C. Marshall, U.S. Chief of Staff in World War II; Jonathan M. Wainwright, hero of Bataan and Corregidor; and George S. Patton, famous for his bold operations in North Africa, France, and Germany.

During WWII Fort Clark became a prisoner-of-war camp, and when it closed it was sold to Brown & Root Company for salvage and use as a guest ranch. Finally, in 1971, the historic fort became the property of North American Towns of Texas, Inc. and is now an oasis-like resort in arid West Texas.

The source of the huge shade trees and rich green golf courses is Fort Clark Springs, which feed Las Moras Creek. A dam forms a wonderful, large three-hundred-foot swimming pool that is a constant 68 degrees, no matter what time of year you visit Fort Clark. Talk about an invigorating dip!

Las Moras Creek (The Mulberries) is also a favorite spot for fishermen, bird-watchers, hikers, and campers. You'll probably want to stay at the modern motel facilities on the Fort's grounds. However, you can camp or park your RV.

The pool is shaded by oak and pecan trees under which Generals Patton, Wainwright, and Marshall once sat. About 680,000 gallons per hour of pure water gush from the springs, so you won't be bothered by chlorination. If you are training for the Olympics or just like doing long laps, the pool is big enough for the whole team to train at one time.

After you've cooled off (which doesn't take long) stroll over to the Fort Clark Museum and learn about the exciting events that took place here during the fort's early days and about its contribution to Texas history. The Medal of Honor winners are buried just west of Brackettville in the Seminole-Negro Cemetery. The museum exhibits a fascinating story of these legendary scouts.

Enjoy a meal in the restaurant which was once the officer's club. The food is excellent. Many of the fort's original buildings have been converted into present-day use, but since they are still intact, you have the feeling of being back in the 1800s.

HAMILTON POOL

HILL COUNTRY

512-264-2740

Location: North on Texas 71 from Austin to FM 3238, north on 3238

Admission Fee

Hours: 9:00 a.m.–6:00 p.m.

Amenities: Parking for 100 cars only, portable rest rooms

It is beautiful at Hamilton Pool. A hundred-foot waterfall cascades over a deep grotto into a sparkling pool, creating a lovely bit of natural beauty in the Texas Hill Country. This pool, with its clear turquoise water and grotto, was formed when the dome of an underground river collapsed thousands of years ago. Early Texas inhabitants found Hamilton Pool and left cultural remains dating back nearly 6,000 years. No doubt the grotto, with its lush plants and abundant wildlife, was a source of great relief from the harsh scrubby landscape.

No one knows what the Indians called this refuge, but the landmark derived its name from Andrew J. Hamilton, tenth governor of the State of Texas. Governor Hamilton was against secession and was voted governor on the Union ticket. He came to the pool often seeking solace and a rejuvenation of spirit. Since then, visitors have continued to come for the same reason. In fact, too many visitors have discovered the fragile beauty of the pool and grotto. Now the county controls the number of people who use the park to prevent overcrowded swimming conditions.

There is no place to change clothes, so come in your swimming gear. Also, it's quite a walk down to the pool, so avoid bringing heavy ice chests and tons of gear. It goes without saying that you need to get here early to be one of the hundred cars allowed. The gates are locked promptly at 6:00 p.m., and if you haven't left the area, your car will be locked inside until 9:00 a.m.

Hamilton Pool is also a hiker's paradise, with trails to explore along the creek that gushes over wonderful holes among the rocks. However, you'll see a number of people sitting in a niche in the rocks and enjoying the water as it pours into the grotto. Just sitting and doing nothing may be the best pastime of all at Hamilton Pool.

KRAUSE SPRINGS

Spicewood 78669
512-693-4181

HILL COUNTRY

Location: On Spur 191 just off Texas Highway 71

Admission Fee

Hours: No set hours

Amenities: Rest rooms, camping, RV hookups

Restrictions: No lifeguard

Texas has a lot of well-kept secrets, and Krause Springs is one of them. To begin with, that's pronounced "KRAU-see." Mr. Krause doesn't advertise his little Shangri-la, but word is getting around, particularly about a spot this beautiful. The National Register of Historic Places knows all about Krause Springs because an undisturbed midden was found here with relics from at least five Indian tribes. Midden is an archaeological euphemism for garbage dump.

Well, the Indians may have been litterbugs, but Mr. Krause's guests take care to keep the pool's beauty intact. An overhanging bluff creates a twenty-five-foot waterfall from the springs above, and the soothing splash of water can calm the most jangled city nerves. Ancient cypress trees drape languidly over the waterfall and lend a somewhat tropical air to the scene. Some of these trees are estimated to be more than a thousand years old.

Very steep metal steps lead down to the pool, which is really quite small for devoted swimmers. Actual swimmable space amounts to a 50-by-250-foot area, most of it near the falls, where the depth approaches eight feet. Because of the numerous boulders in the water, diving is prohibited. What you really want to do is just stretch out on the rocks and emulate a lazy lizard.

However, if swim you must, Krause has built a real sixty-foot swimming pool on the landscaped grounds, and the water is channeled from the springs. Also on the grounds are native stone picnic tables and barbecue pits built by the owner himself. All are shaded by a pecan grove, mesquites, and junipers. Frequent visitors are curious white-tailed deer.

As you drive up to Krause Springs, you may be totally surprised that such a wonderful oasis was once a hog farm. Then you will be delighted that such an idyllic swimming hole still exists in the heart of Texas.

LAKE TEJAS

P.O. Box 673
Colmesneil 75938
409-837-5201

EAST

Location: 1 mile east of Colmesneil on FM 256

Admission Fee

Hours: 9:00 a.m.–7:00 p.m. daily
       May: weekends only
       June 1–Labor Day: daily

Amenities: Lifeguards, camping, RV hookups, bathhouse, rest rooms, concessions

Restrictions: No alcohol

Way back in 1939, some fill dirt was needed for construction of the new Tyler County courthouse. So a kind and generous rancher said, "You can dig the dirt from my property, but the hole you leave has to be in the shape of the state of Texas." Well, they dug, and they carried out the terms of the contract. Springs filled the hole with water, and Lake Tejas was created. Since then, this lake has been an East Texas tradition as a place for a refreshing dip that restores the soul and body from the heat and cloying humidity of the area.

The years have been kind to Lake Tejas. When the rancher died, he left the lake to the Colmesneil (named for a conductor on the Texas and New Orleans Railroad) school with the provision the proceeds from the lake be used to keep it in top-notch condition. Of what value the lake is to the school is quite vague.

Regardless of who owns Lake Tejas, it's a top-drawer swimming hole. Actually, you do most of your swimming in the Panhandle area, but you can bank fish over on the Louisiana border part of the lake. No nasty motor boats are allowed, but Lake Tejas isn't really big enough for motors anyway.

A slide entertains the kids, and the sunbathers slurp up the tanning lotion on three docks, while the loafers stretch out in lawn chairs under ancient pecan and oak trees. You can rent a tube, or it's okay to bring your own float. For the adventurous, a fifteen-foot-high diving board is rather a challenge, and the fainthearted can go off at a short five-foot board into the lake's cool water. However, a word of note—the water gets fairly warm during the hot summer months.

When the rancher said, "Let there be Lake Tejas," he also said, "Let there be white sand." Of course, there is no way to find white sand in Tyler County, so the closest source was Florida. Incredibly, the Florida sand has stayed on its Tejas shores through all these years, and looks as if it will stay around a lot longer. Imagine it! You can be a Florida beach bum in East Texas.

Clarence and Linda Dubose, who run this unique lake, say, "People and things are here today and gone tomorrow, but we'll be here today and tomorrow at Lake Tejas." Let's hope you will be there, too.

LITTLE ARKANSAS (LIL ARK)

HILL COUNTRY

512-847-2767

Location: 7 miles east of Wimberley on County Road 174. Go east on Ranch Road 3237 on the outskirts of town. At the outdoor theater, turn right onto County Road 173. Then follow the hand-painted signs (watch for the right turn onto County Road 174) for several miles of paved and then unimproved dirt road, crossing the river three times before you get there.

Admission Fee

Amenities: Primitive rest rooms, camping, cabins, RV hookups

Restrictions: No lifeguard

After you finally find your way out in the boonies to Lil Ark, expect to be greeted by a crusty old lady named Liza Howell (and a pit bull) who runs her "resort" with an iron fist. You won't believe the list of no-nos posted on the gate: "Don't Dig Worms," "Don't Chop on Trees," "No Dust," "No Motorcycles" (only an idiot would ride a motorcycle on this road). Well, if you think all those signs deter guests, you're wrong. Lil Ark is jammed in spite of "No Electric Toasters or Skillets."

Actually, all the nos are necessary because Little Ark's ecosystem is fragile. Footpaths have left the limestone brittle where several springs run down a heavily wooded hillside.

A small concrete dam crosses the Blanco River, and it is delicious to sit under the dam and watch all your problems float right down the river. The more active crowd loves hanging onto the rope swing and seeing how far they can swing out before letting go. This is about the only spot deep enough for actual swimming. But who cares? Bring your float and enjoy the clear blue-green river without feeling compelled to exercise.

Except for the Japanese cars, the campground does look rather like a place the Okies stopped off in *Grapes of Wrath*. All facilties are really basic, and the granny who greeted you doesn't care. She couldn't stand much more business even if she wanted it. When you pay your fee she sizes you up, and if you don't meet her standards (no one is quite sure what they are, but swinging singles never pass muster) you will be turned away. Hopefully, you'll pass the test, for Lil Ark really is unique.

Making a Splash: Texas Water Parks

On a hot sizzling day, wouldn't it be nice to go to the beach and let the surf toss you around and beat all the kinks out? Or how about just goofing off in an inner tube and not caring where the current takes you? But the closest beach is hours away, and packing all that gear is such an effort.

What's the absolutely perfect solution? It's a water park, of course! And, there's the added attraction of returning home without nitty-gritty sand stuck in uncomfortable parts of your body. In fact, everyone gets in the car to go home sparkling clean. The slides have exhausted the kids, the tots have been wonderful in the kiddie area, and you relaxed as you cruised the park in a tube. You let the big wave hit you a couple of times for water therapy while the nonathletic members of the group poured on the sunscreen and watched from a comfortable lawn chair.

Lunch was just what everyone wanted, served without sand at a shady table. No one had to lug in a heavy cooler. Everyone just went to the concession stand and then munched on nachos and pizza with an ice-cold drink.

Water parks post their safety rules everywhere, and plenty of lifeguards see that they are strictly enforced. Therefore, accidents are practically unheard of. Those monster slides may look scary, but you can bet no one ever gets hurt.

Rest rooms are always clean, and have plenty of room to change clothes. They sure beat the port-a-cans on the beach. Lockers are available for a minimal charge if you want to safeguard your gear. And, if you forgot your sunscreen, just check the gift shop. Then hit the slides or hit the lounge chairs and enjoy it all.

Most water parks offer reduced rates after 5:00 p.m., and all parks offer special group rates. Late in the afternoon is a really fine time to go to the water parks. The crowds have slowed down, and lines are much shorter to get on the slides. You don't have to worry about the killer sun cooking your skin, and the whole park is even more delightful.

Who said that life's a beach? A well-organized water park is the way to go!

Texas Water Parks

FAME CITY WATERWORKS

SOUTHEAST

13700 Beechnut
Houston 77083
713-530-FAME

Location: One mile east of Highway 6

Admission Fee

Hours: May and September—Weekends
June, July, August—daily
Weekdays: 10:00 a.m.–8:00 p.m.
Sat.–Sun.: 10:00 a.m.–10:00 p.m.
Special rates after 5:00 p.m.

Amenities: Dressing rooms, lockers, free parking, lifeguards, security patrol, free life jackets, concessions

Restrictions: No tubes, rafts, or ice coolers brought into the park

Adjacent to the world-class Fame City indoor amusement park, is the equally world-class Fame City Waterworks. Here's a tropical paradise on ten acres of the coolest, wettest water you've ever stuck a toe in. One look at all those sheer-terror slides, and you think, "Not me!" Oh, come on, go for it! Look at all those kids on "Big Mo" as they drop six stories through twisting, turning rapids. They just hop out, head back up the steps, and they're off again. You have to remember that no matter how scary some of those slides look, nobody ever gets hurt on them. Your nerves may quiver for a few hours, but there'll be no breaks or bruises.

The ideal time for Fame City Waterworks is during the week when it isn't so crowded. Weekends often produce long lines to go twirling and bumping down the slides, but

Making a Splash

the kids don't seem to mind the waiting. The thrills are even better with the additional anticipation. A real favorite is the killer that drops three hundred feet fast-forward into the water at speeds up to forty miles per hour. Wow! Now, that's fast!

The kiddie area is very tame with a rain tree, slides, water maze, and kiddie car wash. The huge shallow pool for splashing and giggling is perfect for little people.

Always popular is the Big Wave. The first time you go in, it's a surprise to be hit by a wave of fresh water. Beaches have surf with salt water. But you adjust, it's great fun, and you never have to worry about an attack by an irate jellyfish. And, of course, Jaws wouldn't be caught dead in such hokey surroundings.

The Lazy River circles the park, so catch a tube and join the other floaters taking it easy. Rafts have to be rented, but the tubes are free. You won't find the ambience of the Guadalupe River on your float trip, but the water is just as refreshing.

Lots of concessions serve all the fast food you can eat, and the Heat Waves gift shop has everything you may have left at home. Stake out a lounge chair for some sunbathing and lounging, then hit the slides. Or, should that be hit the Waterworks?

RED CLOUD WATER PARK

Route 6, Box 221
Silsbee 77656
409-385-7858

SOUTHEAST

Location: Take Highway 96 north from Beaumont. Bypass Silsbee to Evandale.

Admission Fee

Hours: May—Weekends only
June, July, August—daily
10:00 a.m.–7:00 p.m.

Amenities: Water slides, picnic area, volleyball, horseshoes, video games, lifeguards, rest rooms, free parking, pool tables, dressing rooms, coolers allowed, tube rentals, concessions

Red Cloud is so laid back, you have almost a feeling of being out in the Piney Woods of East Texas all by yourself. This eight-acre spring-fed lake is surrounded by the tall pine trees that make East Texas so famous. With twenty acres of picnic sites, you really do have a spot all your own.

A nice sandy beach slopes down to the lake, and while the water isn't icy or brisk, it is refreshing. Swimming areas are carefully roped off, and the little guys have a special pool to call their own. The older kids like the slides. Don't expect those monster creations at the big sophisticated water parks, but these slow slides are fun, too. Look at it this way—you don't have to climb all those stairs to slide down.

Popular at East Texas pools is the hand-held "trolley." Kids climb up a ladder and hang on to a sort of trapeze. It travels slowly down a wire, and when it's over open water—let go! Red Cloud has its version of the "trolley," and the younger group really gets a kick out of racing down the wire and falling into the water below.

Picnic tables are back under the trees away from all the water sports. And big groups have their outings in a special area on the opposite side of the lake. You will find the park clean, well-run, and nicely landscaped. The video games are over by the concession stand, so if you came out for some relaxation, you really can find peace and quiet at Red Cloud.

SCHLITTERBAHN

400 N. Liberty
New Braunfels 78130
512-629-3910

HILL COUNTRY

Admission Fee

Hours: May and Sept.—Weekends
Memorial Day through Labor Day—daily
10:00 a.m. daily; closing hours vary

Amenities: Lifeguards, concessions, lockers, dressing rooms, gift shop, free parking
Coolers allowed, but no glass or alcohol

Don't bother to look up Schlitterbahn in your German dictionary: it won't be there. You can translate it anyway you want, and it will always mean wet and wonderful summer fun. Texas has its Wet 'n Wilds, its WaterWorld, and then there's Schlitterbahn. No chain corporation designed this unique water playground. Bob Henry of Houston moved to New Braunfels, conceived of Schlitterbahn, designed it, and today his kids run it. In fact, a great deal of Schlitterbahn's charm is that it's not a chain park, and another delight of Schlitterbahn is its setting on the beautiful spring-fed Comal River.

Pumps installed at the far end of Schlitterbahn lift spring water to the top of the hill behind the castle, allowing the water to flow through a variety of inner-tube chutes, slides, and pools, and then back into the Comal River three thousand feet upstream. As a result, you have refreshing rides in healthy, unchlorinated water.

The minute Schlitterbahn's gates open, the screaming begins and the fun never ends. Where should you go first? If

Making a Splash

you are an old water-park addict, start with the number-one thrill—the Schlittercoaster. You can see this tummy-twister as you enter the park. Get a special sled, climb those steep stairs, and place your sled on the slide's runner. It's too late to back out now, so swallow that lump in your throat, hang on tight, and—CLANK! You are zooming down the Schlittercoaster! Splat! You hit the pool, and now the trick is to keep your equilibrium on the sled as it glides the length of the pool. You manage to do it, or the audience roars with delight as you flip awkwardly into the water. Are you going back for more? You bet! The Schlittercoaster is a challenge to master.

Oh, so you're a novice at water park expertise? Start with the Tunnel Tube Chutes. When you find out how easy and exciting they are, try your skill at the Cliffhanger. It follows a natural cliff, but a word of caution is in order. At a crucial split, you can take a left which ends in a white-knuckle drop-off, or opt for a right turn and ride lazily through the adjacent Landa Resort. After this scenic tour, you hit the river for the trip back to the park.

Wasn't that fun! Now you are ready for the chute of chutes—the Whitewater. Just to be safe, wear extra padding. Cutoffs are allowed, and so are tennis shoes. You are literally rocketed for 1,600 feet through twisting turns and hairpin spins and almost tossed over the edge of the chute. You know you are safe, but you can't help being scared. The newest attraction is the Banzai Pipeline. Double forty-foot-tall tubes of twisting fiberglass descend to the water in spiraling two-hundred-fifty-foot-long twin rides. Oh, wow! Now which are you going to ride again and again and again?

Like all water parks, Schlitterbahn is teenage heaven for girl- and boy-watching. There's also a kiddie pool for future

Whitewater Chute riders who splash away or stop long enough to watch the clown shows.

You can enter the park at reduced rates if you just want to soak in the hot tub, swim in the lagoon, kick a paddle boat, or sit around and watch people having a good time.

The landscaping at Schlitterbahn is very different from that of chain water parks. Unlike those stuck out on a barren prairie, Schlitterbahn is absolutely gorgeous with its beautiful flowers and plants, tree-covered walks and picnic area, and lush hillside landscaping. You feel that you are in the middle of a lovely garden. No wonder Schlitterbahn is a Texas tradition for water lovers of all ages.

Keep in mind that parking is a real problem at Schlitterbahn; get there early.

Next door to Schlitterbahn is the Henry's Landa Resort with a restaurant, hot tubs, swimming pools, tennis courts, and motel-type rooms. Guests receive a discount on tickets at Schlitterbahn. During the season, reservations are a must. Call 512-625-5510, or write Landa Resort, 305 W. Austin St., New Braunfels 78130. A word of note: Landa Resort is totally family-oriented, so don't expect a lot of peace and quiet. Also, the rooms have survived quite a few years of kids, so be prepared for some basic accommodations.

SPLASHTOWN

3600 Pan Am Expressway
San Antonio 78219
512-227-1400

HILL COUNTRY

Location: Five minutes from downtown San Antonio on I-35 North, Exit 160

Admission Fee (discount after 5:00 p.m.)

Hours: Weekends May and Sept. 11:00 a.m.–9:00 p.m.
Daily June–Aug. 11:00 a.m.–9:00 p.m.

Amenities: Lifeguards, concessions, lockers, dressing rooms, volleyball, free parking, raft rentals, gift shop

Restrictions: No coolers allowed, no children under forty-eight inches tall allowed on slides

San Antonio's Splashtown is almost a carbon copy of the Splashtown USA in Houston, so you know you'll have a great time. Formerly Water Park USA, Splashtown has undergone more than a $2 million remodeling, which includes four major water slides, new shade structures, and a totally new children's activity area featuring attractions just for the little Splashtownians under forty-eight inches tall.

Splashtown's enormous wave pool, The Wild Wave, is the best surf action you'll find this side of the Gulf. The surf's always up as four-foot waves give swimmers a real thrill.

Want to take a brief siesta and float your cares away? Just dump your tube in the Siesta Del Rio and drift for a quarter of a mile through its gentle winding course. It's not San

Antonio's famous River Walk, but it's still great fun.

Just as in Splashtown USA in Houston, you'll find The Blue Lagoon with all kinds of water activities, and Shot Gun Falls "fires" its brave riders through a mid-air drop into 10.5 feet of water. And little folks will love the Jumping Platforms and Lily Pad Walk, and Kids Kove with its unique activity center that's guaranteed to keep the kids doused with watery amusement. Nearby is The Dunes. This sand playground is the home of The Wonderfully Whimsical Whizz Worms—two giant multicolored slides standing thirty-seven feet high.

From a sky-high launch pad, riders of the Radical Rampage skip over the water like a stone for more than 120 feet on their own hydroplaning surf-boggan. Talk about thrills! There's nothing like the Rampage for miles.

And, of course, it's here, too. *The Hydra!* It makes all other slides look like a Tinkertoy set. Weaknesicles (that's Weak-nee-si-clees) and Twisticus stretch out more than 272 feet each. Their twin brothers, Chillus and Thrillus, feature two twin loop hydro-spirals. Screamicles and Longdropicus are double-humped speed slides falling nearly 173 feet at a thirty-six degree angle. Who needs the Indianapolis Speedway with Screamicles and Longdropicus?

Pepsi is the official drink at the Splashtowns, and the Pepsi Pavilion is perfect for group outings.

There's great food at every turn. Burgers, hot dogs, pizza, and Mexican dishes are just a few of the goodies on the menu. It's hard to get hot and sweaty at a water park, but you can cool down with Good Humor brand ice cream and other frozen treats.

Making a Splash

Splashtown is San Antonio's premier waterland for the entire family. Here's the place for the most daring "hot-dogger" to the most laid-back "shade-hogger."

Texas Water Parks

SPLASHTOWN USA

P.O. Box 2929
Spring 77383
713-350-4848

SOUTHEAST

Location: 22 miles north of Houston at Interstate 45. Exit Holzwarth Road (No. 68)

Admission Fee

Hours: Weekends only May and September
10:00 a.m.–6:00 p.m.
Daily June, July, August
10:00 a.m.–10:00 p.m.

Amenities: Lifeguards, concessions, gift shop, free parking, tube rentals, dance club, video games, live entertainment, special events, lockers, dressing rooms, rest rooms, picnic area adjacent to water park

Restrictions: No coolers allowed. Children must be forty-eight inches tall for all slides.

Once upon a time many centuries ago, the ancient Greeks were threatened by a horrible monster called the Hydra. This terrible freak had nine heads and the power to grow two more in its place if one were cut off. If all that misery wasn't bad enough, one head was immortal and could never be killed. Finally, the powerful Hercules was sentenced to destroy the Hydra. Hercules solved the problem quite simply. Each time he cut off a head, he burned the stump of its neck. As for the immortal head, he just stuck it under a large rock. So, be careful when you look under rocks—something hideous may be lurking there.

Making a Splash

Splashtown USA has created its own monster called The Hydra. And it is scary. It only has three heads, but it is indeed a Herculean feat of "free-falling" bravery. Towering sixty-five feet into the Texas sky, each gigantic head has its own terrifying personality.

Only the bravest Hydra-maniac will dare to confront Screamicles with its massive height and sheer drop. Challengers of Longdropicus and Weaknesicles will become Hydra-sterical as they speed wildly down the monster's tentacles. Even Hercules' teeth would chatter when he saw this Hydra.

The Blue Beast, while it doesn't strike total terror in your heart, is also a ride to be reckoned with. This sixty-four-foot behemoth of fiberglass and steel has four totally enclosed, blue translucent flumes that envelop riders in a wet whirling adventure.

Rifle down the double-barreled Shot Gun Falls, then plunge into the dark recesses of The Black Hole. Then, it's time to "shoot the curl" of The Wild Wave, Splashtown USA's enormous wave pool filled with more than 500,000 gallons of rockin'-and-rollin' surf. Now, that's fun!

The Blue Lagoon, Splashtown's unique multilevel activity pool, has slides for riding, ropes for swinging, rocks for diving, obstacles for crossing, and challenges for everyone. And, you can count on a lazy ride as you drift along Crystal Creek. It's more fun with a tube, but you can go with the flow without one.

Kids Kove is loaded with pint-sized excitement for all little Splashtownians under forty-eight inches tall. Friendly toads, turtles, 'gators, and lily pads offer lots of fun and frolic for little guys. Be sure to bring the camera Mom!

Splashtown USA fills the summer with fun and music. Live bands perform, and then they stage a gigantic spectacular laser show. Don't miss that zinger! "Beetlefest" brings dozens of custom VW's for a real "Beetle-Bash!" And, if all that isn't enough, come see the Mr. and Miss Splashtown Contest. Call the park for a schedule, as dates vary each season.

More than $7 million were spent in remodeling this lovely water park, and you'll find thirty acres of wild, wet, non-stop fun for the whole family. Pack up the gang, and have a wonderful time at Splashtown USA.

SUMMERFUN USA

P.O. Box 763
1410 Old Waco Road
Belton 76513
Phone: 817-939-0366

CENTRAL

Location: From North: Take I-35 to the Midway Exit. Go south on Old Waco Road.
From South: Take I-35 to the 294-B Exit. Follow signs to Heritage Park.

Admission Fee

Hours: May, weekends only
June, July, August daily
September, weekends only
10:00 a.m.–8:00 p.m.
Sunday: 12:30 p.m.–8:30 p.m.

Amenities: Lifeguards, gift shop, concessions, dressing rooms, rest rooms, free parking, picnic area, tube rentals, coolers allowed

Restrictions: No pets, no smoking, no alcoholic beverages, no glass containers

Sparkling like a little gem in Central Texas is SummerFun USA. Not only is there a gigantic swimming pool, but all sorts of other good old soaking-wet fun is here. The pool is filled with lots of reptilian creatures: twenty-one-foot sneaky snake, a frog that would make Kermit sit up in amazement, a wily alligator, and huge lily pads. No one seems afraid of all those colorful animals. Kids just want to slide down their backs, rest on their scales, or hang on to their tails. But this exotic menagerie does add a real touch of pizzazz to the pool.

For the adventurous, fly like the wind down one of the three turning-and-twisting water slides that take you nine hundred feet before your crash landing in the water below. It will "wet" your appetite (pardon the pun) to climb the stairs and ride the chutes over and over again.

One of the tube slides will dump you right into the Lazy River for a relaxing float around the park. Drop off a minute and pick up an icy snow cone to complete your river trip.

SummerFun has provided several shady picnic areas; the best are down on the banks of the Leon River at the edge of the park. Sorry, no swimming in the river, though. Get your group together and choose sides for a fast game of volleyball. SummerFun has provided the sand for the court, so you don't need a beach after all.

At the Log Cabin you can satisfy your craving for those summer goodies, ice cream and hot dogs. Over at the Smoke House, don't miss the shish kebab on a stick.

Plans are underway for a wave pool to be added and other water park slides. But, SummerFun is a wonderful park already—clean, safe, and cleverly designed. As they say at the Park, "It's not just a sprinkle of fun—it's a splash of excitement for the whole family!"

Making a Splash

WaterWorld

9001 Kirby Drive
(See parking instructions)
Houston 77054
713-799-1234

SOUTHEAST

Admission Fee

Hours: May: Open weekends
Memorial Day to Labor Day: Open daily at 10:00 a.m.
Closing times vary

Amenities: Lifeguards, concessions, lockers, dressing rooms, gift shops
Park at AstroWorld parking lot, located off 610 between Fannin and Kirby, and take the tram to WaterWorld Junction.

First, the marvel of the century, the Astrodome. Next, a bigtime amusement park, AstroWorld. Then, in 1983, WaterWorld. Now this gigantic complex includes entertainment for everyone. How about a baseball game with the Astros hitting those home runs, and then a tram ride across the freeway to the fun and games of AstroWorld, then a few hours on the super-duper slides and in the pools of WaterWorld? There's a day for you. In fact, whoa! Back up! Plan on several days or you won't get it all in.

To take the edge off your nerves, or maybe add a new edge, climb the steps to the top of The Edge. Lie down, fold your arms over your chest, take a deep breath, and Zowie! Kazam! at breakneck speeds you plunge down this eighty-foot free-fall slide. This may be as close to flying without wings as you'll ever get.

For more heart-in-your-throat thrills, head for the Hurricane, Typhoon, Tidal Wave, and Pipeline. These four body slides have several different variations of spirals and corkscrews that twist and turn riders down four hundred-foot chutes over the reservoir of AstroWorld's Thunder River. The Pipeline is a real devil with little portholes here and there to give a tracer-light effect as you zoom through this tube of madness.

With a name like Wipe-Out, you already know what to expect, but why not? This three-hundred-foot slide is aimed at speed, speed to see just how fast your body can go. Ever dream you could get up to forty miles per hour? You can on Wipe-Out, even if it's just for a few quivering seconds.

Want to shoot-the-chute and run the rapids? Sure, you do. Grab a tube and hit the white water in Run-A-Way River. That last shot through the trough before hitting the splash pool is a humdinger. Don't miss this thrill!

For the cowards in the crowd, WaterWorld has waterfalls, a wave pool, and even a nice place for some hand-over-water skills in The Lagoon. Main Stream is for those who love lackadaisical tubing, and the tots will go bonkers with all those clever playground equipment games at Squirt's Splash. You can even try your skills at nonaquatic miniature golf.

Adjoining all this grand stuff to do are several good hotels. Marriott at 800-228-9290 offers some specials and so does the AstroVillage Hotel Complex at 800-231-2360 or 800-392-4398. If you haven't eaten enough in the parks, several

Making a Splash

restaurants are nearby. Kapan's is a tried-and-true seafood place, and it's tops.

For a world of fun in Houston, there's no place like WaterWorld.

WET 'N WILD

1800 E. Lamar Blvd.
Arlington 76006
817-265-3013

NORTH

Location: Just off I-30 across from Six Flags

AND

12715 LBJ
Garland 75041
214-840-0600

Location: Just off 635 at Northwest Highway

Admission Fee (discounts available)

Hours: Mid–May 10:00 a.m.–6:00 p.m.
June–Aug. 10:00 a.m.–9:00 p.m. weekdays
10:00 a.m.–10:00 p.m. Sat.–Sun.
Sept. 10:00 a.m.–6:00 p.m.

Amenities: Lifeguards, lockers, showers, rest rooms, concessions, gift shop, picnic area, free parking, coolers allowed

Restrictions: No glass containers or alcoholic beverages

When it comes to chain water parks, it's hard to beat Wet 'n Wild. The folks who put them together thought of everything that people of all ages love about water parks. If granny wants to come watch the grandkids play in the giant water playground, she'll find reasons to take jillions of pictures. But you don't have to be a tot to get a kick out of the playground. The unique aquatic contraptions are hilarious, no matter how old you are.

Making a Splash

If teenagers want chills and spills (without getting hurt, of course), point them to the Corkscrew Flume. From sixty feet up, they'll spin through a giant figure eight, then drop straight down into a hundred-foot-long mystery tunnel with spray jets and strobe racing lights, and finally splash down in the pool below. Wow!

So the Corkscrew is too tame? Well, climb up the six-story, three-hundred-foot-long Kamikaze Waterslide, the first of its kind in this country. Your body hits a straight shot down a water chute as long as a football field. Talk about bringing the old tummy up in your throat—the Kamikaze is to die for.

So the Kamikaze left you feeling rather ho-hum about water slides? That's hard to believe, but Wet 'n Wild has another monster treat waiting for you—Der Stuka! Like the WWII German dive bomber, Der Stuka drops almost straight down. At a seventy-degree angle, riders "free fall" seventy-six feet—a height sixteen feet taller than the throat-catching Kamikaze Slide. The lines aren't as long waiting for Der Stuka, but what the heck—no guts, no glory.

Those who don't want to climb all those stairs and slide back to nice comfy ground level can do back flips or just swim leisurely through a cascading fountain at the Lagoon Pool. In the center of this beautiful pool is a Pepsi Pavilion for a sip of your favorite soft drink and a sandwich. Or, grab a tube and float down the Lazy River. (Regardless of what the song says, you can't go "Up A Lazy River.")

Wet 'n Wild has a lot more slides to tempt the strong of heart, plenty of shaded picnic areas, and the ever-popular Surf Lagoon. Since it's just across from Six Flags Over

Texas, you can spend days enjoying Arlington. If you stay at the Sheraton in the complex, you can take their free shuttle to both theme parks. The Sheraton has a great pool and hot tub for relaxing after your sightseeing. Another excellent close-by hotel is the Radisson, which offers all sorts of extra amenities for their guests.

Call the Arlington Convention and Visitors Bureau for all the information, 800-772-5371 Texas, 800-433-5374 nationwide.

A Beach Within Reach: Texas Beaches

Texas has 624 miles of tidewater coast line. Marshes and salt grass stretch inland for miles, and mighty petrochemical complexes and refineries dominate a goodly portion of Texas's shores. But 624 miles is a long way, and there's plenty of sand and surf left for everyone who enjoys being a beach bum. In fact, over five million visitors each year make the Texas Coast one of the state's major tourist attractions.

The three "biggie" beaches are Galveston, Surfside, and South Padre, and coming up fast to make it the "Big Four" is Mustang Island near Port Aransas. Smaller beaches like Gilchrist, Quintana, and Matagorda Island are definitely out of the high rent district, and while they have their devoted fans who prefer their beaches somewhat lonely, all of the tourist attractions and crowds are found at the popular beaches.

When you cross the Bolivar Ferry from Galveston to the Bolivar Peninsula, you notice the difference immediately. Gone are all the big hotels and condominiums. All you have available are a few small motels and fishing camps. Many of the houses are for rent, however, but they are not as nice as those on West Beach of Galveston Island. For junk lovers, you will find Bryan Wolf's Trading Post in Gilchrist interesting—if he's in the mood to open.

Two good restaurants over on Bolivar Peninsula are the Stingaree Marina and Restaurant and Zeb's Seafood and Crab House, both in Crystal Beach. However, all in all, there's not much to do. A word of warning: on weekends the Bolivar Ferry is jampacked. Cars have been known to wait hours to cross.

Quintana is a fisherman's beach and rather difficult to get to. No accommodations nor restaurants are found on Quintana. Matagorda is a very nice beach, and a few homes and condominiums have been built, but there are absolutely no tourist attractions. Rather beat-up fishing camps line the road down to the beach, and that is all there is. Matagorda means "fat grasses" in Spanish and has an aura about it that absolutely nothing ever happens here. Stephen F. Austin gave grants to fifty-two families in the area, but the storm of 1894 destroyed the town, and everyone moved to Bay City.

Some wonderful escapes on the coast have no beach at all. You can't beat sitting on the "penthouse" roof of the ancient Luther Hotel in Palacios staring at nothing over empty calm Palacios Bay. Palacios translates as "the palaces," and the historic old Luther is truly a palace in its own way with its special charm (512-972-2312). Forget about a good romp in the surf, though, and swimming isn't any fun either. All is not lost, however. A long fishing pier stretches out in front of the Luther, and you will probably have it all to yourself.

Rockport has just built a small man-made beach, and Corpus Christi has some beach area, but most of the action is at North Padre Island and Mustang Island. Most beach bums agree that the fun of going to the beach is people-watching, and North Padre and Mustang Islands have plenty of that sport to offer.

Texans really get their dander up when "foreigners" say, "Oh, Texas beaches can't compare to Florida or California." We *know* that, but Texas beaches are great in their own way, and people who live on them, or close by, feel sorry for those landlocked folks who can't make a jaunt to the beach just an easy outing. Besides, Texas beaches have their own beauty with spectacular sunsets and pink and purple sunrises, strange seashells, dolphins playing in the waves, and endless whitecaps that never stop rolling in on the sand.

GALVESTON

GULF COAST—NORTH

Galveston Convention
and Visitors Bureau
2100 Seawall Boulevard
Galveston 77550
800-351-4236 (Texas)
800-351-4237 (Nationwide)

Amenities: Bathhouses, lifeguards, surfboard rentals, camping, RV hookups, showers, concessions

Galveston's namesake, the Spanish Count Bernado de Galvez, never visited the island while he was governor of Louisiana. Why should he? Who wants to stay at a salt grass island with ferocious mosquitoes? And the Karankawa Indians were disgruntled hosts.

Jean Lafitte made Galveston his headquarters for a while, and according to the legend, buried his treasure somewhere on the island. The United States grew weary of this pirate raiding its ships, so Lafitte was told in strong terms to vacate his city and take his riffraff with him. Lafitte's final demise remains a mystery.

After the Civil War, Galveston boomed. Magnificent mansions were built, the Strand business district became "The Wall Street of the South," and it seemed that Galveston was destined to become the major city of Texas and the Gulf Coast, and the third-largest port in the nation. All of that was before 1900.

On September 8, 1900, Galveston suffered one of the worst disasters in the history of the United States. A storm struck with a vengeance unparalleled in hurricane history. Carla,

Alicia, and all the rest were just thunderstorms in comparison.

By 4:00 p.m. on the day of the storm, the entire city was under several feet of water, and by 8:00 p.m. the wind was ripping Galveston to shreds at 120 miles an hour. So strong were the gales that a massive tidal wave swept across the doomed city. By midnight, half of Galveston was wiped into oblivion, and an estimated six thousand people had perished. This incredible hurricane is always referred to as The Great Storm, and many buildings and events in Galveston are dated "before The Great Storm," or "after The Great Storm."

The aftermath of The Great Storm was almost as bad as the storm itself, yet Galveston had no intention of becoming another ghost town like Indianola (which never recovered from hurricane devastation). No storm was going to scare Galvestonians from their homes. Slowly they rebuilt. Granite from Marble Falls was hauled in for the seawall, and the city was raised to thirteen feet above sea level. By 1910 more than two thousand buildings had been jacked up, and mud and soil pumped under their foundations.

Fortunately, a few gorgeous homes were spared the wrath of the storm, so Galveston has retained its marvelous historic flavor. Today, the city attracts thousands of tourists. Some come just for the history, others just for the beach, but all leave with a warm feeling for this wonderful old city.

The Coney Island of Galveston is Stewart Beach at the juncture of Broadway and Seawall Boulevards (409-765-5023). You can always count on lots of bodies covering the sand. With a carnival, water slide, bumper cars, go-carts, and a colossal miniature golf course, this beach is an all-

Texas Beaches

time favorite with kids. Complete beach facilities include a large bathhouse and lifeguards.

The seawall is protected from the waves by massive blocks of granite piled against it, and it is a mobile playground for skateboards, roller skates, bicycles-built-for-two, bicycles-built-for-one, and surreys peddled by leg power. Everyone on wheels jostles for room to maneuver. Of course, it's fun just to stroll the seawall and people-watch. You may have to dodge some sun worshippers who bring their cots and lie out for the rays, but the seawall is where everybody does their own thing.

The big sport on Galveston's beaches is to "hang ten" on a surfboard. Well, let's say, the big sport for young strong bodies is to "hang ten." Most surfers have their own special board, but if the sport is new to you, plenty of places have them for rent. You can even get printed instructions from the Visitors Bureau on how to master the art. They make it sound like a piece of cake. So, when the "surf's up," go for it!

Along the seawall there's only a tiny sliver of beach, but if Stewart Beach isn't your style, there's still plenty of island left. It is illegal to drive on Galveston beaches; however, there are cut-throughs from County Road 3005 to parking areas near the surf. You'll find many rules beginning with "no," and they are well enforced, so you'd best abide by them. But, in spite of all those "nos," Galveston's beaches are very popular. Expect a big crowd every day during the summer months. Parking is free.

Campers like Galveston Island State Park, probably because of the screened shelters. Mosquitoes aren't too dreadful when the wind is blowing, but they lurk in the dunes

A Beach within Reach

ready to strike when the sea breeze calms. There are also nature trails and bird viewing blinds in the park, as Galveston Bay is a haven for many species of shore birds. The park is just a few miles out County Road 3305 west of Galveston (409-737-1222). A phone call is advisable to line up one of those screened shelters, as they have only ten of them.

Many campers just park anywhere on the beach and pitch their tents or park their RVs, so if the parks are filled, you can still spend the night on the beach. If you are exploring the world in an RV, the nicest RV parks are at West Beach.

Many Galveston guests prefer to rent a beach house, and plenty of nice homes are in the rental pool on West Beach. Some are downright luxurious and expensive. You'll find a few cafes on this part of the island and some convenience stores, but West Beach is mainly large beach-house subdivisions. The posh condos are closer to town, and the Visitors Bureau has all the rental information including rates, and tons of information on Galveston's myriad attractions. Start your visit with a call to these friendly people and get a thick booklet of discount coupons along with the brochures.

A word of advice: Plan to leave Bowser at home. He has to stay on a leash on the beach, and he's not welcome in the rental houses and condos. Besides, most Bowsers don't know how to get the tar off their feet and will track it into the living room. Also, a lot of Bowsers think that the beach was made for them to roll in dead fish.

For those who want a hotel on the beach, there's the historic old Galvez, now a modernized Marriott, but still holding onto its charm. Galvestonians built it after The Great Storm to prove to the world they planned to make

Galveston better than ever. The luxury Hotel San Luis is almost new. Its swimming pool and patio area is one of the loveliest spots in Galveston.

Out on the end of a pier is the Flagship Hotel. It was once "the place to stay" in Galveston, and is making a wonderful comeback. It is quite romantic to look out from your balcony and see the surf beneath you rather than in front of you.

If you long to be a member of the jet set, you can rent jet skies and jet boats. If horsing around is for you, several stables can saddle up a gentle steed for you. Don't expect to go galloping madly down the dunes. That's one of those beach "no-nos," and horses are not allowed at all on weekends.

A new popular playground is Galveston's Palm Beach, and it certainly is different from Stewart Beach or Palm Beach, Florida. In fact, this ersatz beach is on a saltwater inlet called Offatt Bayou. All it took was twenty-one million pounds of fine white sand from Florida, a concrete freshwater lagoon, some palm trees, and a little tropical paradise emerged. It has everything you could want on a beach, and about all you are allowed to bring in is your bathing suit. Chairs, umbrellas, and paddleboats are for rent, and you must use the Palm Beach refreshment stand if your throat is parched. A sauna, rest rooms, and showers are available. Parking is free, and plenty of lifeguards look after the children. Admission is charged, and hours are 10:00 a.m. to 5:00 p.m. Call 409-331-7256. Palm Beach is at Moody Gardens adjacent to the Galveston Airport.

A bonus at Palm Beach is a visit to the Garden of Life just a few yards away at the Hope Arena. This young garden is beautiful, and in the tradition of Japanese gardens, it serves

A Beach Within Reach

as an allegory suggestive of human life. Admission is free. Check with Palm Beach for hours.

The list is endless for things to do in Galveston when you aren't on the beach. Drive through the Silk Stocking District to see how rich Galvestonians built their homes before The Great Storm. Tour the incredibly magnificent Bishop's Palace. This huge hunk of red stone is an architectural treasure. The American Institute of Architects has recognized the Bishop's Palace as one of the one hundred outstanding structures built during the first one hundred years of that group's activity. It is on a list of fifteen buildings included in the Archives of the Library of Congress as representative of early American architecture. Just down the street is the entirely different, but equally impressive home, Ashton Villa, which is also open for tours.

You'll love the Railroad Museum and the graceful reminder of Galveston's days as a major port when you see the Tall Ship, *Elissa*. A stroll on The Strand is a must. Galveston's answer to Wall Street is now lined by fine shops and restaurants. For gourmet dining, there's the Wendletrap. And, around the corner is the magnificent Tremont Hotel that was restored from an old warehouse building. Also on The Strand, when you see Colonel Bubbie's sign, "The free world's only surviving genuine government surplus store," go in! If you ever wondered what every nation in the world did with their surplus army gear, the answer is Colonel Bubbie's.

Have a leisurely cruise on *The Colonel*, a paddlewheeler that serves an excellent evening meal. The lights of Galveston Bay are beautiful. Or how about an evening at the magnificently restored Galveston Opera House? The symphony may be tuning up, or perhaps a jazz concert is on the bill.

Texas Beaches

Outdoor dramas that are smash hits are performed in the amphitheater on West Beach from June through August.

For famous Galveston seafood, there's the old traditional Gaido's and Hill's Pier 19. Both are very good. There's usually a wait, but the meal is worth it. The Flagship Restaurant is gaining a reputation for fine dining as well.

Why not take a different route into Galveston rather than the frantic pace of I-45? Take the Texas Limited, the excursion train between Houston and Galveston. The trip starts at Houston's Amtrak/Southern Pacific station and stops at Galveston's Railroad Museum. The nonstop run takes a little over two hours. For ticket information, call 713-522-9090.

Galveston has so much to offer that you'll want to visit again and again. The Queen City of the Gulf is regal indeed, and she treats her guests as members of her royal family.

NORTH PADRE ISLAND

GULF COAST—SOUTH

Padre Island National Seashore
9405 So. Padre Island Drive
Corpus Christi 78418
512-937-2621

Amenities: Camping, showers

Not many years ago, Padre Island was called America's vacation frontier. Even now, it still retains stretches of long empty beaches, for the developed portions of the island are on the north and south ends, with some 110 miles in between. At Malaquite Beach, about fourteen miles down the shore from the causeway to the island from Corpus, is the pavilion of the Padre Island National Seashore, your last contact with civilization for more than a hundred miles.

If you plan on hiking, read the Park Service's pamphlet carefully before you set out. Do you really want to carry enough water to last for 110 miles? There is no potable water. Don't forget you'll have to hitch a ride across Mansfield Channel on a fishing boat. Still, you will have 110 miles of beach to yourself unless you run into another hiker or explorers in four-wheel-drive vehicles. Just stay out of the dunes and grasslands, no matter how you travel, as that's rattlesnake territory.

Padre was built by waves and winds, and the ecosystem is extremely fragile. The strong breezes and tides continually change the appearance of the landscape. In some places, dunes inch their way over the grassy flats, and in others they have been stabilized by the binding roots of plants.

Even this stabilization is delicate, for a major storm could start the dunes moving again. If the dunes erode, so will the island.

Nature has created a work of art on Padre Island. Plants such as senna and croton, railroad vine, and evening primrose struggle for survival on these dunes. More than 350 species of birds love the island. Great blue herons, meadowlarks, and the great horned owl are easily spotted. Diamondback rattlers like it here, too; so be on the lookout for creatures that slither as well as fly.

Most of the primitive part of Padre is endless sand dunes, and the scenery never changes. Nature lovers find this sameness of landscape enchanting, however, and come to hike the beach for its isolation.

The first European to land on the island was Alfonso Alvarez de Pineda in 1519 during the same voyage on which he discovered Corpus Christi. It wasn't long before Padre became a graveyard for ships driven onto its shore by fierce Gulf storms. In 1533, a twenty-ship Spanish treasure fleet ran into a hurricane (that was long before the U.S. Weather Bureau started naming the storms) and many of those loaded galleons wrecked on Padre. Of the three hundred survivors to reach shore, only two ran the gauntlet of the savage Karankawa Indians and made it back to Mexico. You may find some gold doubloons when you are out hiking the island or driving it in your four-wheeler, as the Karankawas had no use for money.

About 1800, Padre Nicholas Balli received a Spanish land grant that included the island that bears his name. It's been Padre Island for so long that almost everyone has forgotten it was originally Padre Balli's Island. Parts of his Santa Cruz

Ranch were in use until 1970, when the island became a national seashore.

Before you reach the isolated part of North Padre, many high-rise condominiums are lined up like tall boxes along the beach front. Here is where Mexicans, Germans, and other foreigners have invested heavily in Texas real estate. Luxury beach hotels, golf courses, and cabin cruisers docked at tropical homes symbolize the plush living on this part of the island. What a contrast to the rest of this long skinny island, which has been left in its totally natural state! Most tourists prefer the comfort of the modern hotels and condominiums with the attractions of Corpus Christi just across the causeway.

Even with the advent of the National Park Service, foreign ownership, and thousands of tourists, North Padre Island is still a frontier vacation land. Its splendid dunes, isolated shores, and primitive beauty provide a retreat from the restless pursuits of society. Hopefully, this unspoiled tranquility will be preserved for the future.

ROCKPORT—
PORT ARANSAS

GULF COAST—SOUTH

Rockport Chamber of Commerce
P.O. Box 1055
Rockport 78382
800-242-0071 (Texas)
800-826-6441 (Nationwide)

AND

Port Aransas Chamber of Commerce
P.O. Box 356
Port Aransas 78373
800-242-3084 (Texas)
800-221-9198 (Nationwide)

Amenities: Boat rentals, horseback riding

The Texas Coast's answer to an art colony is the bayside community of Rockport. With its mottes lining the shore, Rockport does have an aura. Yes, that word is mottes, and you won't find it in *Webster's Unabridged*. But when you see a group of the wind-sculpted live oak trees clumped together, that's a motte. The famous trees are the trademark for Rockport, and they look as though Paul Bunyan took his giant saw and ripped away the tops. Of course, we know it was the eternal gulf wind that formed these living masterpieces.

The beach at Rockport is not a vast expanse of sand and surf, but it is clean and fun. A fee is charged to drive in. The Beach Club at the north end of the ski basin offers aquacycles, paddleboats, surfboards, sailboats, and water rollers. At the south end of the ski basin, at Club Wave, you can rent the fast and flashy wave runners, wave jammers, and jet skis. Club Ski, at the south end of the boat ramp, has a

A Beach within Reach

professional ski boat, ski lessons, Kawasaki mini-boat and hydro-slide. The Beach Club opens at 10:00 a.m.

Next to the beach is Rockport's harbor, which is somewhat small, yet the shrimpers add a real touch of color to the little town. That's okay. Everybody knows that Port Aransas is *the* place for deep-sea fishing, so why enlarge the harbor? Where you see the colorful flags flying is Mom's Bait Stand, and you can charter the *Lucky Day* for that deep-sea trip. It says on the brochure, "Nobody gets seasick." Whooping crane tours are also offered on *The Lucky Day* from November through April (1-800-782-2473).

The best new addition to the harbor area and beach front is the new Texas Maritime Museum. An hour or two here can give you a good background on the importance of the sea to Texas.

But thousands of visitors come to Rockport for its incredible birding territory, the richest in Texas (or anywhere else, for that matter). This town is strictly for the birds—more than five hundred species, to be exact. Its most famous winter residents are those rare, gorgeous whooping cranes. You may not be a birder, but you will love the magnificent endangered whoopers. Each winter more and more whoopers return from the wilds of Canada, but it will be many years before they number even five hundred.

Go birding with Captain Ted Appell (512-729-9589). The four-hour trip leaves at 8:00 a.m. and 1:30 p.m. You will learn much about conservation and the dangers that the whooping cranes and rookeries face. After the cranes leave, the shore birds arrive to set up housekeeping and raise their kids. It's a Cecil B. DeMille production with a cast of thousands! You would think that so many birds would be safe from extinction, but you'll learn that many of them will never be seen again. Storms wash away the nests, but the real danger is from man—as always. If the birds are frightened from their eggs for as little as ten minutes, the eggs are ruined. Just think what fishermen, picnickers, and pleasure cruisers do to those birds! Now a new and really dangerous threat has arisen. People living in the Rockport area go out at night and take thousands of eggs. They consider the eggs a delicacy, not caring that the young are being destroyed. Conservationists and the Audubon Society are working against almost insurmountable odds to protect these beautiful birds from disappearing forever.

If you want to contribute to preserving these waterfowl, contact the Audubon Society in your area for information. Rockport has also published a great booklet, "Birder's Guide To Rockport/Fulton." It gives maps with detailed directions, lovely watercolors of the birds, a checklist, references and resources, and a hot line for rare sightings. A spectacular two hundred species can be seen in one day.

Next door to Rockport is Fulton, site of the fabulous Fulton Mansion. This handsome Italinate mansion is a wonderful showpiece, and an outstanding example of how the very rich lived at the turn of the century. The State of Texas has done a marvelous restoration. It is open to the public for a small fee from Wednesday through Sunday.

Rockport's local artists welcome you with some excellent work at the small gallery next to the harbor. Can you resist an oil of those mottes? Or of a gorgeous whooping crane on the wing?

Rockport has a wide variety of seafood restaurants. Charlotte Plummers, a family restaurant, has been around a long time and serves good traditional seafood, including lots of fried shrimp and red sauce. The most entertaining place to

A Beach within Reach

eat is The Boiling Pot. First, tables are spread with clean (for a little while) butcher paper, and you are given a hammer. Next, you order a pitcher of beer. Are you hungry now? Well, get set, cause here it comes—dumped right on the table in front of you. You get a steaming savory mix of Cajun-seasoned corn, shrimp, crabs, and potatoes, so start banging those crab claws with your hammer and have a memorable feast.

Rockport has numerous condos and motels. Some look like pre-Hurricane Carla, but a brand new one is the highly recommended Laguna Reef (1-800-248-1057). Beautifully appointed rooms with moderate rates are available. They have a private beach, pool, and a thousand-foot-long lighted pier. The suites have a fully equipped kitchen and a wonderful balcony to watch a pink-and-golden sunrise. The beach/lagoon is somewhat rocky, so take your tennis shoes. A free continental breakfast of juice, coffee, sweet rolls, and toast is also part of Laguna Reef's amenities. A restaurant is not available at the hotel. Rates are extremely reasonable, and even lower in the off-season from September through May 1.

Best Western has a fairly new Rockport Rebel Inn (800-528-1234), but it's not on the gulf. Key Allegro is a swank and affluent beach-house community built on canals for easy docking of private boats. Key Allegro is quite pretty, and it has a marina and pool, but no beach. Call 512-729-2333. The Sand Dollar Resort is right out of the fifties. Set in a beautiful live-oak grove, the resort has many units with kitchenettes, and rates are reasonable. It's just across from Captain Ted's departure point for the early morning birding cruise (512-729-2381). In Rockport, life's not so much a beach, as a bird.

Take the free ferry over to Port Aransas and watch the dolphins cavorting in the waves. It's a short trip, though sometimes a long wait behind a line of cars. But, the dolphins never get tired of amusing you, so it's worth it.

Years ago, before condominiums, Port Aransas was *the* place for tarpon fishing, and the old historic Tarpon Inn still has its famous collection of signed tarpon scales. Among the signatures is that of Franklin Delano Roosevelt, who came to Port Aransas and the Tarpon Inn in the thirties.

The Tarpon Inn used to have a wonderful view of the pass as the big ships and tankers came down from Corpus to the open seas. Condos arrived, squelched the view, and in fact just about squelched Port Aransas. Not long ago this tiny community was the only place on the coast that might be called quaint, with a lot of "seaside-tacky stuff." That's all gone. The Tarpon Inn will endure because it is on the National Register, but the once excellent restaurant at the inn is closed, and rooms look like they did when Roosevelt was here.

The place to stay now is down on Mustang Island. It's a lot less crowded than South Padre and every bit as nice, with ultramodern condos right on the beach. Mustang Towers offers all the amenities you could want in a high-rise (512-749-6212), but if you prefer a motel arrangement, try Island Retreat (512-749-6222). Highly recommended is the luxurious Port Royal with a swimming pool worthy of a Hollywood extravaganza. Everything about Port Royal is first class, yet rates are very reasonable. By all means, don't miss the view from the Royal Beachcomber Restaurant at dinner (1-800-242-1034). There are many more condos, and the Chamber of Commerce will send you a list. A word of caution: spring-break kids have discovered Mustang Island, so you may want to time your visit accordingly.

Horseback riding is allowed on the beach, so no matter if you've never been on a horse, Mustang Riding Stables has a slow gentle nag just for you. They even have old-fashioned horse-drawn hayrides with fireside cookouts. Just give them a call at 749-5055 and reserve old Dobbin.

If there is a deep-sea fishing capital of Texas, it is Port Aransas. A new fishing tournament seems to start every weekend. The biggest is the "Deep Sea Roundup" held around July Fourth. Hundreds and hundreds of pounds of kingfish, sailfish, marlin, redfish, and shark are weighed in daily. Surely you want to see your trophy hanging on the scales, so contact Fisherman's Wharf (512-749-5760). If they are booked, lots of other boats are available, and the Chamber has the list.

For that really fine seafood dinner, try the Spaghetti Works. For yummy hamburgers, a crisp shrimp salad, and a great Bloody Mary, you will like the casual Tortuga Bay. On the porch you can still see the big ships slowly chugging out to sea.

Quaint and tacky are passé at Port A (as it is affectionately called), but it still has a charm that you won't want to miss.

GULF COAST—SOUTH

SOUTH PADRE ISLAND

South Padre Island Tourist Bureau
P.O. Box 3500
South Padre Island 78597
512-761-6433
800-343-2368 (Texas)

Amenities: Boat and fishing rentals, camping

For a "tip of Texas" vacation, the place to be is South Padre Island, Texas's answer to Miami Beach. While Miami has its rows and rows of hotels, South Padre has its rows and rows of condos. Most of them look just alike, tall and modern and rather on the posh side. The chain motels and hotels are here, too: Holiday Inn, Hilton, and a lovely new Sheraton. All of the Sheraton's rooms look out on the sea, and furnishings are quite elegant. Yet rates are competitive with the condos, and in the off-season the Sheraton is a real bargain.

To reach this island paradise, you drive across a big new causeway over the Laguna Madre from Port Isabel, and a bronze statue of Padre Balli welcomes you to his former domain. The good padre was given a land grant to this slender stretch of island, and instead of becoming Balli Island, it ended up Padre Island.

Many things on South Padre are new, but Port Isabel has been here for a long, long time. The decrepit old lighthouse was used by both Yanks and Rebs during the Civil War. Stop and climb its twisted staircase, and you can see why troops wanted this lookout point.

A Beach within Reach

You also pass one of the tackiest gift shops in the world. The stuff inside is everything you never wanted made from seashells. It's not every day you see a gigantic plastic octopus waving to you from the roof to come inside. Remember the song from the 1950s about the girl in the "itsy-bitsy, teensy-weensy, yellow polka-dot bikini"? Well, this lovely octopus has on a bikini, but it's an "itsy-bitsy, teensy-weensy, purple polka-dot bikini." Oh, go ahead and get a seashell wind chime, or night light, or ashtray, or . . .

Also in Port Isabel is the historic Yacht Club. Back in the good old days when citrus farmers were very rich, they built an exclusive private club they called the Yacht Club. Even though it has gone through several remodelings, the rooms are still small and dark, but the restaurant serves the best food you'll find on the "tip of Texas." During the season, the Yacht Club dinners are so popular you have to call for a reservation for one of three seatings (512-943-1301).

Now it's time to hit the beach! After all, that's why you're here. The water is clearer at South Padre than at Galveston or even North Padre, but the sand is not any whiter. Surfers like the surf at this end of the island best, and it is definitely the "in place" to be at spring break.

South Padre has everything a beach bum could possibly ask for. The islanders want to make sure you come back over and over again, and you will. How about sail boating or water-skiing on Laguna Madre's placid calm waters? Wind-surfing is great, too. Just be careful not to disturb nesting birds, for hundreds of species make their homes at this lagoon. Rental stores have everything you need for water sports, including lessons.

For those who prefer the company of fish, Padre is building

an artificial reef from old army vehicles. To get rigged out for diving, call Captain John Palmisino at 512-943-5569.

As you are strolling the sands of South Padre, you are sure to run into Walter McDonald, the Frank Lloyd Wright of sand castles. His condos, shopping centers, resorts, and subdivisions are masterpieces in wet sand. You are more than welcome to become one of his construction workers, but you have to furnish your own spade. Blueprints are not required.

The Tourist Bureau really brags about the deep-sea fishing at South Padre. They claim that the big trophy fish are here at South Padre, not Port Aransas. To prove they know their fish, South Padre citizens stage the Texas International Fishing Tournament the first of August, and it is more exciting than any other fishing event in Texas. The state record game fish caught off South Padre Island are blue marlin, bluefin tuna, blackfin tuna, tarpon, Mako shark and wahoo. The wahoo was caught by a rank amateur. Terry Crider from Denton caught his 124-pound trophy on only his second deep-sea outing.

On every block of South Padre T-shirts are for sale, so if you missed a souvenir at the seashell shop, you can browse for T-shirts to take home.

Scampi's Restaurant on the island is run by the same owners as the Yacht Club, so you know that it's good. You might also book a dinner cruise on the paddlewheeler *Isabella Queen* as it putters around Laguna Madre. A buffet is served with cash bar and live music.

Another cruise of a different sort entirely is aboard *Le Mistral*, South Padre's gambling ship.

A Beach within Reach

Sure, you come to South Padre for the beach, but it's also an international playground. You can cross the border to Matamoros to shop for Mexican treasures and dine on excellent food at Garcia's, The Drive-In, or the U.S. Bar. All mix those divine margaritas that are the best in the world. Also, a Garcia's tequila sour is superb.

You can easily drive over the bridge and park at Garcia's, where your car is always under surveillance, but getting back to the U.S. side involves a long, long wait in traffic to clear customs. Your best bet is to just park on the American side and walk over the bridge. Don't forget you must declare your bargains in alcohol when you return, and you're only allowed two bottles per person duty free. You still have to pay Texas tax.

Another must is the Gladys Porter Zoo in Brownsville, one of the best zoos in the country. At Harlingen is the famous Confederate Air Force Museum. What a memorable way to learn about the Air Force in World War II. All of these antique prop-driven planes still fly, and members of the CAF give them a lot of tender-loving care.

Back on South Padre you can meet an incredible woman who has lived an exciting life far ahead of her time. As a young girl, Ila Loetscher joined a women's flying corps headed by Amelia Earhart, and her stories are wonderful. Again marching to a different drummer, Ila led a movement to save the Ridley sea turtle from almost certain extinction. Her crusade earned her the title, "Turtle Lady of South Padre Island."

Ila is in her eighties and still going strong in her effort to save her beloved turtles. She rescues and nurses injured turtles and then returns them to the sea. Those which are too badly hurt and can never return to sea become part of her "family." The turtles have an extensive wardrobe, certainly one *Vogue* would never approve of, but Ila says that the costumes make the turtles more memorable to her audiences and promote saving them from extinction. Meet the Turtle Lady, at 5805 Gulf Boulevard, with her show at 10:00 a.m. Tuesday and Saturday from September through April, and at 9:00 a.m. May through August.

South Padre abounds with legends of buried treasure and lost gold. Not only are Spanish doubloons said to have washed ashore, but Jean Lafitte is thought to have anchored in Laguna Madre. If he tossed any coins overboard, the mud has claimed them forever. One treasure you might dig for is the one the Singers buried. They had purchased land from the good Padre Balli and were known Union sympathizers. When the Civil War winds blew too close, the Singers buried their valuables and left Padre for the safety of Yankeeland. They returned about a year later to find those fickle sands had shifted, and all efforts to locate the treasure were in vain. So get out your shovel (perhaps a metal detector would be better) and start digging. The Singer fortune may be just under your feet.

SURFSIDE BEACH

GULF COAST—NORTH

Surfside Beach Tourist Council
Route 2, Box 9009
Surfside 77541
409-233-7596

Location: Take Hwy. 288 south of Houston to intersection of Hwy. 332; follow Hwy. 332 east 6 miles to Surfside Beach.

Amenities: Fishing rentals, camping, cabin rentals

Back in the fifties Surfside Beach was nothing but a few ratty houses on pilings and miles and miles of salt grass. Horses and cattle chomped on the grass, and if you were asleep in one of the cabins, you might awaken with the sensation that Surfside had miraculously become an earthquake zone. Those sturdy pilings made wonderful scratching posts for Bossy and her buddies.

Lots on Surfside sold for $100 each or two for $150, and the only access was a rinky-dink drawbridge across the Intracoastal Canal. Barges had the right-of-way, so any trip to the beach necessitated carrying along a good book and mosquito repellent. You could spend hours (no exaggeration) waiting for the drawbridge to descend so that the road became a road again.

The massive petrochemical complex, Dow Chemical, dominated the landscape, but east toward San Luis Pass were some great stretches of lonely beach. Even in midsummer with the beach at the height of its popularity, you could still spread your blanket on a spot to call your own.

Well, the good old days are long gone. Surfside is now one of the most swinging beaches on the coast.

Dow Chemical hasn't gone anywhere. In fact, it has grown larger and larger. The cattle and horses have moved to greener pastures, and lot prices have jumped to four and five figures, but there's also good news—the drawbridge was replaced by a modern bridge that looks high enough for the QE2 to cruise under.

Discos, motels, condos, and beer joints are all frantically busy, and Surfside is unique in that you can still drive the sands without barricades. Every variety of vehicle shows up to see and be seen in. Surfside is definitely a young crowd's beach.

Plenty of nice beach houses are available for rent, but the ratty ones are mixed in with the nice houses, and the overall effect of Surfside is somewhat seedy. Certainly, none of the posh condos of Galveston, Mustang Island, or South Padre make the Surfside scene. Lots of rental agencies are in business here, and the Tourist Council has all of their brochures. Everything for light housekeeping is furnished except linens.

What would a fishing port like Freeport be without deep-sea fishing? In fact, most groups prefer to depart from Freeport rather than from Galveston. Action Charters (409-265-0999) can fix you up for an exciting all-day excursion in search of monster game fish. But check around. Some other outfit may have a trip that suits you better. The Tourist Council sends *lots* of information for you to peruse. Just don't forget the dramamine.

If deep-sea fishing is too steep for your pocketbook or too hard on your stomach, long jetties are popular fishing spots. A lot of big fish have been pulled in from these jetties.

Even though the beach has a bunch of cafes, you don't want

to miss the Old River Inn restaurant on Second Street in Freeport (409-233-0503). The menu is mostly seafood, and with it comes some wonderful iron-skillet cornbread. Your talented waiters will flip it over in the skillet for your entertainment. Does it ever land on the floor? Of course. Nobody's perfect. But don't worry, there's lots more skillets in the kitchen. Dress is casual and reservations are not required.

Another popular seafood restaurant off the beaten path on County Road 519 in Brazoria is Dido's. With a beautiful view of the San Bernard River and good food, there's usually a wait. The ads for Dido's have a map, but you can always call 409-964-3167 for directions. Dress is casual and so is the service, but it's always fun.

As you are heading for the high bridge to Surfside, you'll pass Ernie's. No, don't pass Ernie's. You simply have to go in and be overwhelmed by all the stuff Ernie's can sell you for the beach. It matters not what you forgot, Ernie's has it—plus gobs more. Also, it's a great place to ask directions to Old River Inn and Dido's.

You don't go to Surfside to tour the area as you would in Galveston. Old Velasco was the site of a battle during the Texas Revolution, but what there was of the town is now incorporated into Freeport. You come to Surfside for sun, sand, and surf. Isn't that what the beach is all about?

How to Be a Texas River Rat:
Texas Rivers

Nueces, Frio, Blanco, Colorado, Guadalupe, Brazos, Neches, Sabinal, Trinity, Sabine, the mighty Rio Grande, and Red River are all names that appear over and over again in the story of Texas. The many Spanish names are a legacy left behind by early conquistadors and priests—a legacy that will endure as long as the rivers flow.

Boundaries were set and fought over where the rivers cut their way to the sea. After the revolution against Mexico, Texas claimed the land north of the Rio Grande as far west as Santa Fe. Then Mexico claimed Texas as far north as the Nueces River. Peace treaties didn't carry a guarantee of peace, and it took years to establish the present boundaries of Texas. We don't think much of those rip-roaring days unless a history teacher insists we learn about them. Today Texas rivers are for fun. They may flood occasionally, but soon another dam will be built to take care of that problem.

A lot of Texas rivers just drift along with nobody noticing them, but others are famous vacation traditions. What kind of summer is it without a tube trip down the

Guadalupe at New Braunfels? Or a yearly getaway to Neal's Lodges on the Frio? Or being one of the bumper-to-bumper crowd at Garner State Park? If you are a dedicated rafter, you may spend all year working on an entry in Waco's Great Raft Race on the Brazos each Labor Day. And scores of other special events test the skill of devoted river rats.

Of all Texas's many rivers, those in the Hill Country are perhaps the most scenic with their limestone cliffs and pristine waters. But consider the majestic canyons of the Rio Grande. For absolute grandeur, these massive sheer walls of rock cannot be surpassed. A Rio Grande raft trip is a Texas must.

The best way to explore a river is in a canoe, but canoeing is a real art. It requires a lot of experience to take on an entire river. Special sections of rivers have been set aside as ideal canoe trips that even a novice can easily handle, so begin with those. But no matter how lazy a river can be, please wear a life jacket. Leave the white water to the pros until you get perfect control of the canoe.

Many Texas rivers are running bodies of water filled with mud. The Red River got its name from the tons of red dirt that washes from its banks. As the Colorado and Brazos empty into the Gulf of Mexico, they too become a muddy brown color with much of their beauty destroyed by soil and debris. So most of the recreational areas of the rivers are far upstream.

Most river rats have a favorite river. Each is so different, and all are fun to explore. Some rivers you will enjoy for their exciting history. The Red River could tell about the wild and wooly days of cattle drives, and the Sabine about its gory days of the "Neutral Strip" that became a no man's land for every criminal type in the country. The

San Jacinto could fight the famous battle, and the Rio Grande could keep you spellbound with stories of Texas Rangers and Mexican bandits.

So, pick a river, any river, go with the flow, and become a Texas River Rat.

BIG THICKET

SOUTHEAST

Nick Rhodes, Canoe Program
Big Thicket Museum
P.O. Box 315
Saratoga 77585
409-274-5892

OR

Big Thicket National Preserve
8185 Eastex Freeway
Beaumont, Texas 77708
409-839-2689

Just what is the Big Thicket? It got its name when early migrants from Louisiana on their way to Texas found that their way was effectively blocked by impenetrable thickets from the Sabine River on the east to the Brazos River on the west. The Big Thicket covered more than two million acres, but today less than twenty percent of it remains.

To the scientist, the Big Thicket is an area of great plant and animal diversity produced from complex interactions between soil formations, drainage, and climate. They see a giant field laboratory in which many natural events can be studied.

To the layman, the Big Thicket is a place with a special quality, mood, and soul. It has resisted human intrusion until recently, and many bitterly resent any intrusion at all. Lumber companies have ripped vast areas to shreds and continue to do so, replanting the stripped areas only with pine trees rather than natural vegetation. Devastation has been so rampant that in the 1960s the Big Thicket Association was formed to save the little that was left. Finally,

How to Be a Texas River Rat

in October of 1974, 84,550 acres became the Big Thicket National Preserve. Composed of twelve separate units dispersed within a fifty-mile square, only a few areas are available for use.

To really get back to nature in the Big Thicket, take a fascinating canoe trip with Nick Rhodes down Village Creek. You start at 9 a.m. and arrive at the takeout point about 5 p.m. Nick will send you all the information, and the minimal cost goes to the upkeep of the Big Thicket Museum which needs financial help so badly.

You start at the rustic museum with Nick's slide show of the wildlife you might see on your trip. Because Nick's a science teacher, you'll learn that nonpoisonous snakes have round eyes and poisonous snakes have slanted eyes—except for the deadly round-eyed coral snake. Come on, Nick, who's going to stop and examine a snake's eyes? You run!

Village Creek runs through an area of the Big Thicket that belongs to a private organization known as the Nature Conservancy. Its goal is preservation of the nation's natural beauty, and its pet project in Texas is the Big Thicket. Nick tells you how tons and tons of garbage were removed from this heavily wooded area so it could return to its natural state and you could canoe in a nature lover's paradise.

Village Creek is spring-fed, and the water cold and clean. You have to maneuver around a lot of logjams, some caused by those sharp beaver teeth, so watch carefully where you paddle. Cypress knees hug the bank looking like a colony of little people waiting for a ride. Nick says there are two thoughts as to why cypress trees have knees. One is they stick out of the water so the roots can breathe, and another is that they are stabilizers to keep the trees erect. But, the local theory is that cypress knees are there for you to trip over.

You soon pass Photographer's Bend. It wasn't named for its photogenic qualities, but because a Fort Worth photographer dropped his camera gear in the creek at that spot. At a deserted railroad bridge Nick pulls in for lunch and a two-mile nature hike. The snow-white sand beaches along the creek are really lovely. At every bend is a perfect picnic spot and swimming hole.

As you tromp through the woods on an unmarked trail, you learn about all the plants and soils that excite ecologists. You go from a flood plain to a cypress slough to a desert within a few hundred feet. Nick points out the Dolly Parton tree with its unusual nodules, and then the infamous hanging tree. You also learn to identify spiderwort, larkspur, snake cotton, and green-eyed Susans. Nick will also show you the strange little racerunners. They are all females and reproduce through parthenogenesis. Talk about women's lib! These critters can run thirty miles per hour.

The hike takes you past Alligator Lake; those big guys are hard to spot among the jillions of water lily pads, but take your binoculars just in case. You won't find any 'gators in Village Creek, nor have the ravenous nutria found their way here.

After lunch (you bring your own), you paddle up a dark mysterious slough filled with slimy green water. Trees cling precariously to the bank, and some have given up the ghost and crashed into the water. Stark white egrets stand out magnificiently in these murky surroundings. The whole scene is breathtaking.

How to Be a Texas River Rat

You may not get to see any of the four species of carnivorous plants in the Big Thicket, and you may miss the rare and delicate orchids. You must remember that these plants are highly endangered, and all sorts of red tape is required for guides to point them out. Plant thieves may return and dig them up for sale. Also, Nick has learned to keep to himself all of his animal sightings. Hunting is permitted in this part of the Thicket, and a bullet may be waiting for that special animal Nick spotted.

You really hate to turn in your paddle when it's time to drag in the canoe. What a special part of Texas you have shared, and you are so glad at least some of it is left for us to enjoy, thanks to people like Nike Rhodes who volunteer their spare time to the conservation of the Big Thicket.

BRAZOS RIVER

NORTH CENTRAL

Rochelle's Canoe Rental
Route 1, Box 119
Graford, 76045
817-659-2581

Amenities: Camping, canoe rental, concessions, picnic area

The Brazos is a great river for a novice canoeist. Not all 840 miles of the river Spanish explorers called *Brazos de Dios*, or arms of God, are recommended, but the twenty-mile stretch between Texas Highway 16 and Highway 4 in Palo Pinto County is definitely the most scenic. "It loops and coils snakishly from the Possum Kingdom dam down between the rough low mountains of the Palo Pinto country," said John Graves in his novel *Goodbye to a River*, the story of the author's three-week canoe trip down the Brazos in which he tells exciting tales about the Kiowas, Comanches, and white men who lived along its shores.

According to one legend, a band of Indians atacked a Spanish mission near the river and killed everyone but one friar and a few Indian converts. Frantically the group crossed the river, and the warriors spotted them. As the screaming Indians rushed across the river, a great wall of water caught them in midstream and swept them away. The friar was so grateful, he proclaimed they had been saved from torture and death by the arms of God—*los brazos de Dios*.

North of Possum Kingdom Lake the river is a dreary, dirty brown, but below Possum Kingdom the water is quite clear and a refreshing sea-green color. Towering cliffs and limestone boulders are covered with trees every Texan will recognize—cottonwood, oak, mesquite, cedar, and pecan. All along the banks are shady, deserted sandbars that are perfect for a lazy picnic and a cool swim. Don't expect any white water; the Brazos is for the indolent canoeist who likes scenery and doesn't have to worry too much about getting his gear wet. The river teams with fish, and wildlife is everywhere along its banks. Life is very easy along this stretch of the "Arms of God."

Rochelle's can accommodate just about any length trip you want from one-day outings to overnight camping. If you take the one-day, get there early. Rochelle's has lots of canoes, but it takes a long day to finish your jaunt. *Most important: Call in advance and get them to give you an honest answer on the depth of the water!* You can easily spend eight miles of the ten-mile trip dragging your canoe, getting angry and having a miserable day because you didn't find out the water level. Insist on a truthful answer, and if the water is low, *don't go!* You'll find Rochelle's on the north side of the Dark Valley bridge on Texas Highway 4 between Graford and Palo Pinto. You'll be shuttled upriver for a ten-mile canoe trip back to Rochelle's. Flying along the shuttle ride on those curving gravel roads in a free-wheeling pick-up is more exciting than the actual canoe trip.

Many canoeists like to take the overnight camping trip, and even inexperienced canoeists can easily make the trip in two days. Or you can camp at Rochelle's after the one-day trip. However, if you do the whole route from Highway 16 to Highway 180, plan on four days.

If you forget something, you can bet Rochelle's has it—even plastic garbage bags for your gear at twenty-five cents each. Try not to forget anything; Rochelle's is the only store in miles and prices are high.

A deposit is required on each canoe, and shuttle service is extra. The only things you don't pay for are the life vests and paddles. Get a price list from Rochelle's before you even leave home. A word of caution: Even in May that Texas sun is a killer.

With its scenic banks, slow clean water, and great fishing, the Brazos is the perfect river for learning the art of canoeing and is highly recommended for large groups on a canoe outing.

COMAL RIVER

HILL COUNTRY

Chamber of Commerce
P.O. Box 311417
New Braunfels 78131-1417
512-625-2385
Reservations hot line
for New Braunfels: 800-545-6606

THE TUBE CHUTE
512-625-4251

Admission Fee

Amenities: Rest rooms, tube rentals

Restrictions: No lifeguard

CAMP WARNECKE
512-625-3710

Admission Fee

Amenities: Rest rooms, bathhouse, cabins, concessions, video games, tube rentals, camping

Restrictions: No lifeguard

In Texas almost everything is the biggest or the best. In New Braunfels they brag about the shortest: the world's shortest river, the Comal. The Comal may be short (three and a half miles), but it is fueled by the largest springs in the

Southwest. Shortest, largest, and sweetest just about describes the Comal River.

If you ever wondered where old inner tubes go to die, the answer is New Braunfels. Renting "toobs" may easily be the town's leading industry, particularly in the summer, and business is always booming. The big-time favorite tubing spot on the Comal is the Tube Chute built by the city to chill and thrill all ages. This S-shaped concrete chute can really be scary and somewhat dangerous because of the strong current. If you love your feet, wear your sneakers.

Float down to Camp Warnecke, which has been there almost as long as the Comal. This ancient family resort is surrounded by new modern condominiums, and Camp Warnecke has joined them. Nary a vacancy is to be found on weekends, and not many during the week. Call 800-292-1167 (Texas only), and hope something is available, because these really are the best on the river. Everything you expect in a quality condominium is here, and many have woodburning fireplaces, washers, and dryers. The condominium complex also has a private swimming pool, clubhouse, and hot tub. More casual accommodations are available in the motel units which also overlook the Comal. In the pavilion you will find a snack bar and tube rental stand. For somewhat more rustic accommodations, try The Other Place at 512-625-5114.

A small dam at Warnecke creates a few bumpy rapids that float you along at a rapid pace and make the upstream return easy.

If you come to the Comal to swim and see that sea of "toobers," don't despair. Between Landa Park and Prince

Solms Park (the Tube Chute) is Hinman Island City Park. The World's Shortest River becomes about thirty feet wide and ten feet deep and stays a constant 71 degrees. A straight quarter-mile stretch can be easily swum. But, on weekends, forget it. The only way for peace and solace is to hold your nose and dive deep. Just rent a tube and join the gang at Warnecke.

In 1987 the Guadalupe went on a rampage, and the Comal remained calm, cool, and collected. Of course it was also jammed to the brim with "toobers" who couldn't get on the Guadalupe. Most years as the summer wanes, the water drops to a trickle in the Guadalupe, but for those who still want to go tubing, there's always the constant-level Comal.

If you weren't lucky enough to get a condo on the river, I-35 is lined with chain motels, and the T-Bar-M Ranch, just two miles from loop 337 on Hwy. 46 to Boerne (512-625-7738), is highly recommended. With its tennis courts, condos, and activities for kids to do, this is a super base for enjoying New Braunfels.

All that fun in the sun is guaranteed to make you ravenous. Don't cook every meal in your condo; get out and savor that good German kraut, schnitzel, and wurst. Krause's Cafe is downtown at 148 South Castell and serves hefty portions of German and American food. New Braunfels' natives keep Krause's tables full.

Oma's Haus, 541 Texas 46, is very popular with tourists, as is the Smokehouse at the intersections of Texas 46 and Texas 81. For a quieter atmosphere, the Log Cabin on I-35 just south of town serves excellent German dishes. (See the section on the Guadalupe River for more information.)

New Braunfels has really promoted its rivers and produced the hottest cool spots in Texas. No summer is complete without at least one New Braunfels weekend, preferably many weekends. Here are some definite Texas "must do" attractions.

GUADALUPE RIVER

HILL COUNTRY

Chamber of Commerce
P.O. Box 180
New Braunfels 78130
512-625-2385
800-445-2323 (Texas Only)

Amenities: Tube and canoe rentals, camping, picnic area, rest rooms

When it's "toobing" time in Texas, *the* place to ride the rapids is the Guadalupe River between Canyon Dam and New Braunfels. This section of the river's flow is determined by the amount of water released by the dam, and the volume is measured in cubic feet per second, which river rats always refer to as "cfs."

The cfs is posted for the public, but even a novice who never heard of cfs can figure out when the river is at a dangerous stage. For sheer excitement, the flow is at its best above the 600 cfs mark, but it is not a stream for a beginner, small children, or anyone on the outside of too much firewater. Your best bet is to go with a very experienced guide. It will be a river trip you'll never forget.

On most days, the Guadalupe is an easy-flowing, gentle river with bank-to-bank floaters in every imaginable sort of getup. Sensible types have on T-shirts over bathing suits, tight-fitting hats, sturdy tennis shoes, and sunscreen lotion within easy reach. Lots of shade covers the water from the old trees hanging on the banks, but it doesn't take much sun to make you look like a Maine lobster just fished from boiling water and ready for the drawn butter. Tennies protect your tender feet from those slick, sharp rocks if you go tumbling over in one of the rapids. Or the river may be running so slowly that you have to drag your tube over shallow water. That's tedious and more on the order of work.

Late in the summer, if the rain has been nonexistent, the Guadalupe slows down to a trickle. A three-hour tube ride can take all day, and some years this miserable river condition lasts all summer. Your best bet is to call the Chamber or one of the canoe rental places and ask how the river is running.

"Toob," raft, kayak, and canoe rental places line the river, and you really need a reservation for all but the tube. Old tubers *always* get tubes with a bottom. They cost extra, but that piece of plywood can save a very tender portion of your body from being ripped to shreds. Everybody in the group gets a tube with a bottom, plus an extra one just for the ice chest. That is almost as big a necessity as tennies. Please, *no glass*. Then, everyone ties their tubes together so they won't drift far apart, bites the bullet, and leaps in the brisk, spring-fed water. It's only cold for a few minutes. As soon as you hit the hot sunshine, that cold water feels very good indeed. Now you are ready for the rapids. Canoeists and kayakers need a very early start on the river, or they will be caught in a logjam of tubers. Everybody has a great time, and it is almost a tradition that everybody in Texas gets in at least one tube trip on the Guadalupe at New Braunfels.

Naturally, the speed and excitement of the rapids depends on the water level of the river, but it's all fun as you bounce round and round, up and down, bump off of rocks, dodge tree limbs and other tubers. Some groups form a line with their legs locked on the tube in front, and while this may not be as graceful or rhythmic as the conga line, it brings screams to a fever pitch. It's a thrill not to miss. Just don't lose the ice chest.

The rental places shuttle you upriver and drop you off for either a three-hour or a five-hour float. The price is about the same, so just decide how long you want to stay encapsulated in your tube. You pay a rental fee, a shuttle fee, and a deposit per tube. In some places they take cash only, so come with money. Canoes, raft, and kayaks are more.

Parking is a major problem along River Road. Whatever you do *don't* park in no parking zones, and *don't* park on the road. These laws are strictly enforced, and they will tow you away before you can even close the car door. The best time to hit the river and the River Road is early in the morning, particularly on weekends. You can easily get a tube with a bottom and a parking place. You can leave your keys with the rental company, and please do—or you may lose them forever to the rocks and mud of the Guadalupe River.

As for food, a few beer joints and barbecue places are stuck between the private homes that line one side of the river. Best to ask the outfitter how to look for them. Don't try to snack in your tube unless you really hanker for soaked bread and dripping brownies.

A really great time to be on the river is late Sunday afternoon. All the campers have gone except for a few who have extra days to loaf, and the floaters have gotten their deposits back, leaving the river so quiet you can actually hear the birds. It is a real pleasure to have this world almost to yourself. The Guadalupe is a fantastic river, and sometimes the ambience is lost in the mob scene. Yet, the mob scene is part of the Guadalupe, too, and all the friendly floaters add to the fun.

Another real problem on the Guadalupe (and other rivers) is garbage. Some of it is unintentional, but it's nevertheless there, creating an eyesore on this big tourist attraction. Please put the empties back in the ice chest, and if you see some refuse floating by, pick it up. Every fall a river cleanup is held. If you want to help keep Texas rivers beautiful, call 512-625-2385.

Camps line the Guadalupe, and on weekends the campers really have to love togetherness—with adjoining campers. Usually the camps rent on a first-come, first-serve basis, and by late Friday night, you can forget it. Even the chain motels that line the interstate are often booked on weekends. The river and the other great New Braunfels attractions really pack the town.

The most romantic inn in New Braunfels is the Prince Solms, and the best food in town is in its basement at Wolfgang's Keller (512-625-9169). Another fine place to stay is in the quaint hamlet of Gruene (pronounced Green) at the Gruene Mansion Inn (512-629-2641). You'll definitely want to go "kicker" dancing at Gruene Hall and join the crowd waiting to eat at the unique Grist Mill Restaurant. Unlike Wolfgang's Keller, which is a white-tablecloth establishment, the Grist Mill has no dress code for its hamburgers and chicken-fried steak.

Contact the New Braunfels Chamber of Commerce for a list of river outfitters and campgrounds, plus all the numerous other things to do, and the area will give you a big hearty "Wilkomen" with a coupon booklet saving you a lot of money on restaurants, motels, and attractions.

(To reach the River Road, take Highway 46 west toward Boerne, look for a blinking light, and turn north when you see the street sign. For more places to rest water-soaked bodies, see the section on the Comal River.)

GUADALUPE RIVER STATE PARK

HILL COUNTRY

Park Road 31
Route 2, Box 2087
Bulverde 78163
512-438-2656

Location: 8 miles west of the intersection of U.S. 281 and Texas 46

Admission Fee

Amenities: Camping, rest rooms, showers, RV hookups, picnic area, swimming, hiking

Restrictions: No lifeguard

For a year or so Guadalupe River State Park was hardly known to anyone but Texas Parks and Wildlife, but it was just too good to keep secret, and now it's one of the state's most popular parks. No wonder. It offers the scenic Guadalupe River, camping, tubing, canoeing, swimming, and hiking, but none of the commercialism of the River Road below Canyon Dam. Sure, it can be crowded, and getting a campsite may be only a dream, but its natural beauty isn't hidden behind a line of drying bathing suits and cars parked end to end.

The park's most outstanding natural features are huge bald cypress trees, two steep limestone bluffs, and four natural rapids. Best of all for canoeists, there is a place to land their canoes.

The Guadalupe River is classified as a navigable stream, and according to Texas law it is open to anyone who wants to float down its length. The river may be open, but its banks are privately owned, and landings are only with the permission of the owner. As you can imagine, it would be a rare property owner who would let canoes find a resting place among his livestock. At the state park, canoes can glide up to the banks, pull ashore, and avoid irate ranchers who have no use for canoists.

Multi-use campsites are nestled in the woods, and primitive, walk-in campsites are near the bank and picnic area. Nature trails offer wonderful hikes, but in the summer most park visitors head for that cool, emerald green river.

PEDERNALES FALLS STATE PARK

HILL COUNTRY

Route 1, Box 31A
Johnson City 78636
512-868-7304

Location: Seven miles north on FM 3232, off U.S. 290 East of Johnson City

Admission Fee

Amenities: Hiking and nature trails, swimming, tubing, picnic area, camping, rest rooms, reservations recommended

Restrictions: No lifeguard

That's Purr-den-alice, y'all, pronounced the way President Lyndon Johnson did when he lived on this river. It's one of Texas's shortest rivers, less than a hundred miles in length, but here in the park the elevation drops about fifty feet over a distance of three thousand feet, and the falls are formed by the flow of water over the tilted stairstep effect of layered limestones. For you geologists: these limestones belong to the Marble Falls formation of the Pennsylvanian geologic period with an age of over three hundred million years and are the oldest rocks exposed in the park. Since Texas is rather short on tall waterfalls, Pedernales Falls is downright spectacular.

Wandering up the ledges and boulders that form the falls, you can find tadpoles as big as your thumb and impressions in the rocks that resemble footprints and even dinosaur tracks. Wildlife is very abundant in the park, so watch for the whitetails, and in the fall a bald eagle may just glide swiftly overhead. Why not get a bird list at headquarters and mark off your sightings? At least 150 species have been seen in the park. Good luck!

It's quite a steep climb down to the swimming area. Unfortunately, the swimming area is not beneath the falls. Dangerous currents and hazardous water conditions have closed the falls to bathers. Still, the clear green river has a great section for that cool dip or lazy float on your tube. Just follow the signs for the swimming area. It is absolutely worth the hike down. Just travel light, because you have to come back the same way.

Pedernales is one of the state's best parks for scenic beauty and outdoor enjoyment. The state of Texas purchased the land in 1970 for nearly a million dollars. It was a working ranch and is dotted with seeps and springs and dissected by intermittent streams and canyons. Unfortunately, the Pedernales is also subject to flash flooding and rising water. The campsites are safely on high ground, but if you hear a siren while you are along the river, *leave the area immediately as fast as you can.* This happens only once in a blue moon, but don't ignore the flash flood siren if it sounds.

You are in LBJ Country at Pedernales Falls, so why not visit the Lyndon B. Johnson boyhood home in Johnson City, just nine miles away? Twenty-eight miles west at Stonewall, the Lyndon B. Johnson State Historical Park and National Historic Site offers daily bus tours.

RIO FRIO

SOUTHWEST

Southwest Texas Visitors' Guide
Uvalde Leader-News
P.O. Box 740
Uvalde 78802
(Enclose $1.35 for postage)

Amenities: Camping, cabin rentals, rest rooms, showers, concessions

The Frio River rises in Leakey (that's Lake-y) and its surroundings are spectacular. The west bank butts against a high limestone outcropping, and juniper, oak, mesquite, and cypress line both banks. Frio is Spanish for cold, and during hot, dry Texas summers, cold can feel mighty good. Actually, the gorgeous bluish-green water in the Frio stays in the seventies all year long, which is just right for dipping. Small dams along the river's path have formed top-drawer swimming holes, and even though this may be thought of as a fairly remote section of Texas, the world has discovered the Frio.

Ten thousand people can't be wrong, and that's about how many campers and assorted Frio lovers cram into Garner State Park's abused facilities on weekends. Paddleboats and water trikes vie with floaters for space. As you can imagine, the noise is on the excessive side. Peace and quiet are not the order of the day, and romantic souls can just forget Garner State Park. But for families this is a paradise with facilities for children and a dance terrace for teenagers. The Saturday night dances during the summer months attract about five hundred kids. If there's a live band, watch out! About two thousand teens could show up.

In addition to the Frio, camping, and dances, Garner visitors can enjoy a miniature golf course, hiking, and biking. Just make your reservation early—like a year in advance (512-232-6132).

To enjoy the cool jade waters of the Frio, you certainly aren't limited to Garner State Park. Other camps, some on the rustic side, edge the Frio's cypress-lined banks. Ordering the Southwest Texas Visitors' Guide from Uvalde's *Leader-News* is highly recommended.

Camp Riverview strongly frowns on any blasting radio. It is in a lovely setting with lawns, pecan trees, and a nice gentle tube run (512-232-5412). It's just down the river from Garner and Concan on U.S. 83.

The granddaddy of Frio resorts is Neal's Lodges just off U.S. 83 on Texas 127. Neal's was touted by *Southern Living* magazine as ". . . one of the most scenic swimming holes on the cypress-lined river." It's certainly one of the oldest and has been a family vacation spot since 1927. Some of the original rustic cabins are still available. Other cabins have been added through the years and have various dates. Ask what they have lined up for you, as some still aren't air-conditioned.

Except for a few short years, the Neal family has operated the lodge from the wonderful general store with everything from flea collars to chicken soup. Don't panic if you've forgotten your sunscreen or hairspray; the Neals will find it for you.

Most of the cabins have kitchenettes, but you simply *have* to try Neal's Cafe. It's in one of the original buildings and perches on a cliff overlooking the river. As you might expect, there's old-fashioned chicken-fried steak, plus roast beef and other Texas fare, to be washed down with gallons of iced tea. Breakfast is popular, too, so you might have a wait on weekends.

The swimming hole is nice and deep, and has warm rock islands for basking, and diving boards for showoffs and cannonball leaps. Tubes are for rent or you can bring your own. The water never gets over 78 degrees, and on a 105-degree day that seems almost frigid. Before you start on a tubing jaunt, check with the store on the water level. Sometimes you may end up carrying your tube rather than floating in it.

Most guests at Neal's are lured by the beautiful Frio River, but it's also for nature lovers, birdwatchers, hunters in season, and folks who just don't want to do anything. The active group can ride horses, hike, and wander around the countryside poking in antique stores or the famous Leakey Drug Store.

If you would like your own private house on the river, call LeAnn Walker at 512-232-6633. LeAnn runs the Rio Frio Bed 'N Breakfast and handles rentals.

Fall is gorgeous along the Frio with the reds, yellows, and browns of the foliage. November is best, and the colors often last until after Thanksgiving.

RIO GRANDE   SOUTHWEST
Far Flung Adventures
P.O. Box 31
Terlingua 79852
915-371-2489

There are many ways to experience a river, but the *best* way is to be on it in a canoe, a kayak, or a raft. Not only do you enjoy its scenic beauty, but you learn its currents and personality. Rivers are not just streams of water running to the sea; they shape men's lives and destinies. Wars are fought over their boundaries and civilizations rise and fall on their banks.

The river that has carved so much of Texas's history is the eternal Rio Grande. Nowhere on its course is the scenery more spectacular than in Big Bend National Park, and for a trip you will never forget, Far Flung Adventures will raft you through the Rio Grande's awesome canyons.

It's a long way to Big Bend, and no matter where you live there's lots of lonesome prairie to put behind you before you get there. But it is worth every mile! Indian legends record that the Great Creator had a pile of rubble left over when he finished shaping the heavens and the earth. He threw the rubble into a heap and formed the Big Bend Country. He also formed a geological showplace. Remnants of one hundred million years are exposed by erosion and overwhelm the spectator with what Nature can do when she has the time.

All of the superlatives apply in describing Big Bend. Rare birds and animals still survive in this protected haven, and due to its very remoteness, Big Bend is a remarkably preserved area. Somehow man has resisted changing it to suit a foolish fancy.

Nobody ever described the Rio Grande's waters as pristine. They are ugly red mud, but you don't come here to scuba dive: you are rafting down pages of history of primitive Indians, Spain, Mexico, Texas, and the United States. What stories this river could tell—and what a great job the Far Flung's guides do of interpreting them.

Steve Harris, who started Far Flung Adventures back in 1977, believes that besides being fun and educational, river camping is "therapeutic, an escape toward simple self-reliance, from the dependencies and demands of the electronic age." Equally irresistible are nights under the stars, meals eaten by the campfire, and days of discovering what might be around each bend. And Steve is absolutely right. Lying under the light of billions of stars with the gentle lap of the river to lull you to sleep is the grandest relaxation in the world.

The best times of the year to raft Big Bend are spring and fall. You can count on Easter and Thanksgiving being booked solid everywhere in the park. Winter isn't too bad, as it rarely gets bone-chillingly cold, but summer is a merciless killer in Big Bend. Daytime temperatures are usually over a hundred degrees, and the sun sucks your body dry. No one in their right mind goes to Big Bend in the summer. Plan in advance so you can sign on with your outfitter when the river and the weather are at their best.

The most popular trip is the two-day float through Santa Elena Canyon. Santa Elena is a place where, as one visitor aptly wrote, "If light were sound, this canyon would be a symphony." You float from sun to shadow and back again beneath walls of limestone that sometimes top fifteen hundred feet. Well along in your journey looms the canyon's major navigational hazard—"The Rock Slide." Here the river weaves and threads its way through a jumble of massive boulders the size of big houses. Your guides study The Rock Slide carefully, then you are on your way! For the novice it looks impossible, but as soon as you finish the maze and thrill to the incredible twists and turns, you are ready to go back and do it again and again and again. You feel that you have learned something about the river, and perhaps about yourself.

When you request for information on the trip, you will receive a thorough and beautiful brochure with color photos of what you will see. Nothing has been left out; it tells you what to bring and what is expected of you. Actually, about all you have to bring is your sleeping gear, a plastic drinking cup, and your toothbrush. Far Flung really takes care of you, and not only does it look out for your gear, it serves some of the best campfire chow you have ever tasted. There's never a dull moment and never a dull meal. You are expected to help carry the gear in and out of the rafts and to help set up camp, of course.

The rafts leave from Lajitas early in the morning, and you are back from the trip late afternoon the next day—after a very rough ride through the desert from the end of the float trip. Your guides are experts at rafting and on the moods of the river, and you never have to worry about their decisions. A lot of the trip is spent just drifting along, dipping the paddles occasionally, and maybe having an occasional thought as to what the rest of the world is doing.

What a letdown when the trip is over! You want it to go on and on. But there's always another time, another float. So, pack up these memories and plan on more.

For information on where to stay, contact National Park Concessions, Big Bend National Park, 79834, 915-477-2291.

The Chisos Mountains Lodge has motel-style accommodations, or you can camp. Also, Terlingua has a few motels, and Far Flung can give you the numbers, or you can drive to Lajitas and stay at the modern Cavalry Post Motel or the Officers' Quarters, which are almost-new condos. Call 915-371-2471. A word of caution: Big Bend is a long way from anywhere. Have a reservation before you go.

SABINAL RIVER

HILL COUNTRY

Lost Maples State Natural Area
Utopia 78884
512-966-3413

Amenities: Camping, cabin rental, parking, rest rooms, showers, picnic area

The Sabinal River doesn't have all of the glory of the Frio. It's only fifty-eight miles long and doesn't have those special swimming holes and float trips so dear to a river rat's heart. However, the Sabinal (Spanish for cypress) is wonderfully scenic, and in the past few years some delightful getaways have opened and are giving the Sabinal a claim to fame other than its beauty.

The Sabinal rises up above tiny Vanderpool, near Lost Maples State Natural Area. When you think of maple trees you think of New England, but here in a tiny protected canyon are groves of maples. Every fall the tour buses head for Lost Maples, but nature is fickle, and some years the leaves are rather lackluster. Your best bet is to call the park for information on the ideal time to see the reds, golds, and chestnut browns of this little freak of nature.

Camping is allowed at Lost Maples, but the area is so small that it is advisable to call the park first. Also, parking is a real problem at the peak season, and the canyon entrance may be bumper-to-bumper. To really appreciate the gorgeous colors, hike back into the canyon to see the trees. Another plus in going to Lost Maples is the drive to the area. If the weather cooperates, the Hill Country puts on a real show for fall. The entire countryside is filled with muted autumn shades, and they can last as late as early December.

About three and a half miles north of Vanderpool is Fox Fire (512-966-2200), and here are log cabins that Abraham Lincoln never dreamed of. All are furnished in antiques and country decor and have modern kitchens. They are right on the Sabinal, and since there are only six cabins, you won't have to share your area with a crowd of other river lovers. Fox Fire doesn't offer full resort facilities; and you can forget TV, telephones, and restaurant meals. In fact, it is best to bring all of your own food, as restaurants are sparse.

Utopia on the River advertises as "a riverside bed and breakfast lodge," but actually it is more like a motel, with color TV, a meeting room, jacuzzi, sauna, and a swimming pool. Still, the setting on the Sabinal is very scenic, and a hike should definitely be on the agenda. Breakfast is the only meal served. Call 512-966-2444.

You can find out much more about the Sabinal and its many attractions from the Southwest Texas Visitors' Guide (see the chapter on the Frio River). Also ask for a Wintergarden antique guide. The area has about twenty-one antique shops to explore for treasures. And, don't miss seeing Uvalde's restored Opera House and the John Nance Garner museum.

Utopia was named for Sir Thomas More's book of the same name in which he envisioned the perfect civilization. More probably had much more in mind that tiny Utopia, Texas, but for most visitors, Utopia is truly idyllic. You will find beautiful scenery, peace and quiet, and a total change from big-city life. Now, that's Utopia, Sir Thomas.

SAN MARCOS RIVER

HILL COUNTRY

Tourist Development Council
P.O. Box 2310
San Marcos 78666
512-396-2495

Location: Take the C.M. Allen Parkway exit off I-35, and the river is one minute away—in the heart of San Marcos.

Amenities: Tube and canoe rentals, bed and breakfast inn, camping

You almost forget there is a San Marcos River because of all of the razzmatazz promoting Aquarena Springs. But a fairly large river courses out of those 99,000,000 gallons of water that discharge daily from the springs. The Indians knew about the springs and the river long before the Spanish arrived about two centuries ago. The Spaniards first called it *Rio de los Inocentes,* or river of the innocent. Later explorers christened it San Marcos, or St. Mark. The old conquistadors loved to leave behind names of their patron saints. A mission was established and a fort was built, but neither was successful.

Divers feel you can't really know the San Marcos until you dip your head below the surface of the water with snorkel or scuba gear. The water is crystal clear, and you can identify more than fifty species of fish living in these waters. Some of the common river inhabitants are the sunfish, bluegills, catfish, perch, and gar. No other river contains more than three species of the bass genus, but the San Marcos has four: the largemouth, smallmouth, spotted, and Guadalupe.

Look for the mouthbrooders. These strange fish pack their eggs in their mouths until they hatch. Even after birth, females open their jaws to the young fish when trouble approaches. Another unusual denizen of the San Marcos deep is the freshwater eel. It must journey from the San Marcos to the Gulf and then to the Sargasso Sea in the North Atlantic to mate. It makes the entire year-long trip without feeding.

If you are very lucky you will spot a giant freshwater shrimp. This monster may reach up to a foot in length and weigh as much as three pounds. Males are armed with large crusher claws, and they look formidable indeed. However, they were the victims of a thriving shrimp industry during the early 1900s, and are now so rare that they are almost more mythical than real.

The rarest of aquatic plants, Texas wild rice, lives in the San Marcos and nowhere else. The total growth of the plant could probably fit into a small two bedroom house, and each year the production is less and less.

An underwater archaeological expedition in 1978 uncovered 100,000 artifacts, proving that man's relationship with the river dates back 12,000 years. No one knows what name the Paleo Indians gave the river, but you can bet it was a good one.

At the source of the springs is the beloved Aquarena Springs, an all-time favorite tourist attraction. You can cruise around on those little glass-bottom boats and watch fish and see where one of the archaeological excavations was made. It would be exciting to see one of those three-pound shrimp, but don't count on it.

But, you didn't come to Aquarena Springs just to see a "dig." You came to see Ralph the Swimming Pig do his stuff in the underwater ballet show. Most pigs love mud, but Ralph loves water. He will leap right in, and his short legs will churn the water like crazy—as long as his trainer is in front of him with his yummy bottle of milk. The show is fun and kids love every minute of it. Ralph is available for interviews after the show, but does not give autographs.

There's also the sky ride, the tropical gardens, trained macaws, and a Texana village. You can't swim in the Springs (unless you are a pig named Ralph), but historic Aquarena Springs Inn has an olympic-size pool, and the golfers can tee off at the Aquarena Springs Golf Course.

A movie was filmed at these springs. You may not have caught it at the theater, but watch for it on the late show. *Piranha* stars the springs' killer perch attacking actors in wet suits covered with peanut butter.

Aquarena Springs Inn has been here since 1928 and has run the gamut from school to hospital to health spa. Now it is refurbished and has modern motel-style rooms. A wonderful balcony overlooks the springs, and you'll love watching the glass-bottom boats move slowly across the tiny lake. For reservations, call 512-396-8900.

Now that you've started at the beginning of the river, let's see what else this clear, cool (72 degrees) stream has to offer. Instead of heading down the Interstate to tube at New Braunfels, why not avoid the crowds on the Guadalupe and tube the San Marcos? You can rent tubes and canoes in City Park (no phone), and even on a frantic Sunday you can enjoy a lot of peaceful drifting in your rubber ring of relaxation. There's no white water, but the dense and lush vegetation hugging both sides of the river creates a real feeling of solitude. In fact, you might well be drifting down the Amazon. Just don't open a jar of peanut butter!

If you are a very accomplished canoeist, you may want to contact Spencer's Canoes, Rt. 1, Box 55-R, Martindale 78655. You can find out about joining up for the Texas Water Safari—the world's toughest canoe race, held each year in June. You and a partner (some make it solo) paddle your canoe 260 nonstop miles from San Marcos to Seadrift on the coast. Record time is 35 hours and 26 minutes, but *surely* you can beat that time! In a race of this sort, winning isn't necessarily the goal. It's just finishing the darn thing—who cares how many hours it takes?

Spencer's Canoes has everything a canoeist or kayaker could ever use. They teach both sports using new and innovative techniques that purportedly give you more skills than any other canoe schools. A few sessions with Spencer's and at the next Texas Water Safari you'll be ready to set a new world's record.

For those who just like a lazy trip down a beautiful river, you can do that here too. Spencer's will put you in, take you out, and provide guides. Next to Spencer's is the Shady Grove Campground. Canoeists rate it the best little campground in Texas. It is a canoeing access point, a meeting place for races and special events, and a place to come to watch the river drift by. Leave all the party animals over on the Guadalupe.

Those who want to stay in one of the best bed and breakfast inns in Texas can call Crystal River Inn at 512-396-3739. It's located at 326 West Hopkins, the main street of San Marcos.

San Marcos boasts numerous other tourist attractions and streets lined with wonderful old Victorian mansions. There's always some big event taking place. So, plan to stay around town for several days and really enjoy all San Marcos has to offer.

UPPER GUADALUPE RIVER

HILL COUNTRY

Kerrville Area Chamber of Commerce
1200 Sidney Banker
Kerrville 78028
512-896-1155

Amenities: Boat and canoe rentals, camping, picnic area, bed and breakfast, RV hookups, rest rooms

How about a vacation in one of the healthiest climates in the United States? According to a Rockefeller Foundation study, West Kerr County has it all—mild temperatures with low humidity, cool and dry summers, and winters tempered by warm air from the Gulf. For Texans on the Gulf Coast who often see one hundred percent humidity, that just might sound like heaven on earth.

Lots of people do come to Kerrville and Kerr County for the healthy air and others come for fun and games on the icy waters of the Guadalupe River. The water isn't actually icy, but it is definitely brisk. During a Texas summer, brisk is good.

If you are staying in Kerrville and want the joys of a river vacation, they're available at the Best Western Inn of the Hills (800-292-5690). The old part of this traditional motel is centered around swimming pools, but the luxury condominiums are right on the river with views of paddleboats, sailboats, and canoes leisurely enjoying the river. What are you waiting for? Go down and rent one and get out on the river. Sure, the condos have every amenity you can think of, including a gourmet restaurant and lounge with live entertainment, but don't waste the sunshine and the river just watching. Put on your swimming gear and let the excellent River Club put you on the river. Or how about a romantic canoe ride under a big, yellow Hill Country moon? That can be arranged, too.

Since you can't spend you entire life on the river, the Inn also has golfing, tennis, and a complete sports center.

If you came to the Upper Guadalupe to camp or park your RV, you'll find many choices in the Chamber's Area Accommodations Guide. Don't miss Kerrville State Park. Check out its five hundred acres on Flat Rock Lake before pitching your tent. Call 512-257-5392 for more information.

Lots of funky motels, camps, and parks line the river on Texas 39. Mixed in with the places for big people are the *crème de la crème* of private camps for kids. If you want your kids to have a whale of a summer, this is the place, but it isn't inexpensive. You could probably take a cruise or two for what some of these camps cost.

When you get to Ingram Dam on Texas 39, drag out the picnic basket, open a cold one, and watch the kids slide down the dam, or join them in a slide or two yourself. Just cool off in the swimming hole if sliding isn't your thing. Incidentally, Texas 39 from Kerrville to Hunt and then FM 1340 are a lot of fun. Along every inch of the Guadalupe is a recreation spot, and they come in all sizes and prices.

Down the road a bit from Hunt is Casa Bonita (512-238-4422). Its little gray-and-white boxes are neat but hardly modern. The yard looks as if an army of gardeners works on it, and the putting green is in perfect shape. If you prefer, there's a swimming pool instead of the river. Most guests don't prefer. Isn't the Guadalupe why you're here?

River Bend Bed and Breakfast is located between Camp Stewart and Camp Waldemar. The limestone building is right on the banks of the river, and can accommodate up to thirty guests. Furnished in antiques, River Bend is very popular, so call early for reservations (512-238-4681).

Toward the end of all the camps, motels, and such, you'll find River Inn Resort. It may well be your choice for a Hill Country vacation. The rooms are motel-style with a big family room with the only TV. However, aerial hookups are in the rooms. What you'll love about River Inn is a super deck overlooking the dam and its swimming hole. Grills are available for barbecue cookouts, and tubes for river floating. The flume is on the tame side, but who needs excitement? Rowboats and canoes offer exercise for anyone with lots of energy, and if that's not strenuous enough, you can backhand a few over the nets of the tennis courts. Call 512-238-4226.

Kerrville practically bursts with festivals and events. Get their calendar and plan an outing to suit your taste. Guaranteed to please everybody is the Texas State Arts & Crafts Fair on Memorial Day and the following weekend. Tour the Y.O. Hilton if you like Western art; it has some of the best. The fabulous Cowboy Artists of America Museum is also here, so allot a few hours to see the most talented Western artists in the nation. For jewelry lovers, it was in Kerrville that James Avery originated his popular pieces, and you can visit his shop.

Y.O. Ranch is nearby with its exotic animals and well worth a tour, and so is MO Ranch with its unique and beautiful buildings. The nice people in Kerrville want you to "lose your heart to the hills," and you'll just do that here on the upper reaches of the Guadalupe River.

Lakes Superior: Texas Lakes

Minnesota may boast of being the Land of Lakes, but Texas might well claim that title. Just look at a Texas map—splotches of blue are everywhere (well, with the exception of Odessa and Big Bend). Only Alaska's glacier-locked expanse exceeds the more than 6,000 square miles of inland water in Texas. You have probably never heard of a lot of the lakes, and others are nationally famous, like the Highland Lakes. Get a copy of the State Highway Department's "Lakes Trail" from the Travel and Information Division, P.O. Box 5064, Austin 78763. You will be astounded at the little lakes that never made the big time. Maybe folks want to keep their own special lake quiet so they don't have crowds ruining them.

Visitors do not come from far and wide to visit Lake Whitney, Lake Tawakoni, or Carter Lake. And it is doubtful that many Texans ever heard of Lavon Lake or Benbrook Lake. The state is filled with these small lakes that are off the beaten path. No fabulous marinas line their shores; motels and resorts are not interested in them; but they have a lot of pleasure to offer people who go there. Perhaps it's just a fisherman in his bass boat who wants to get away from the local factory, or maybe the kids think of it as their swimming hole and are glad it's small and unknown.

Many of Texas's lakes are man-made, but who cares? They are just as pretty as if the Lord made them himself. And lakes offer a great deal of enjoyment. You certainly can't water-ski in beach surf. Fishermen like to anchor their boats close to shore and see if they can outsmart those freshwater bass. Other fishermen don't care if they catch anything at all. And lots of people just like to sit and look at a lake and think of absolutely nothing.

A guide to all of Texas's lakes would be about the size of an encyclopedia, but the following are some of the biggest and best the Lone Star State has to offer.

LAKE AMISTAD

Chamber of Commerce
1915 Avenue F
Del Rio 78840
512-775-3551

SOUTHWEST

Amenities: Scuba diving, boat and horse rentals, camping

Restrictions: No lifeguards

Nobody ever said the scenery in the Trans-Pecos region of Texas was beautiful. In fact, the land was so harsh that it was the very last area of Texas to be settled, and it took the railroad to accomplish that almost impossible feat. Some parts of this barren desert are still sparsely populated, but Del Rio is quite a bustling little city. This is probably because so many water enthusiasts and fishermen know about one of Texas's most beautiful lakes. The sparkling, deep blue water of Lake Amistad creates quite a contrast in this land of prickly pear cactus, scruffy mesquite, and throat-choking caliche dust.

For eons the confluence of the Rio Grande, Pecos, and Devils rivers produced devastating floods that destroyed lives and property. Finally, the United States and Mexico agreed to build dams along the Rio Grande, and in 1968 Amistad Dam was completed. Amistad means friendship in Spanish, and this dam and lake symbolize the spirit of friendship between the two nations.

Take a short drive over the dam and see the impressive statues of the lordly eagles that symbolize each country. Note the narrow line in the concrete that is the boundary between the United States and our neighbor to the south. The dam created a lake with fingers reaching into steep, closed canyons that lure fishermen after the big ones, the hybrid "super bass." Clear blue-green water also makes the lake one of the finest diving spots in Texas, particularly in the cooler months from November to April. For full details, call Ramona Wallace of the Amistad Scuba Divers shop, 512-774-6422 or 512-774-4170. Ramona is also a PADI (Professional Association of Diving Instructors) instructor.

As West Texans know, the wind never ceases in this part of the world, so how about getting high on a parasail or trying your skill at tacking in a sailboat? You don't have to bring your own equipment. At American Watersports on Highway 90 you can rent anything that floats. The number is 512-775-6484. In fact, you can also rent a clip-clop, clip-clop-slow-easy saddle horse here. And, if you want to camp, a site can be arranged. That's how versatile Bob Perry is at American Watersports.

Lake Amistad is part of the national park system, and rules are rather strict. Only certified instruction is allowed on the lake, but Godwin Thomas at American Watersports can teach you any water sport you want to learn. This transplanted Englishman can have you water-skiing, sailing, or parasailing in no time at all.

Not far from Del Rio, west on Highway 90, at Seminole State Historical Park, are some of North America's oldest pictographs. About eight thousand years ago, primitive man left his version of graffiti in Fate Bell Canyon. If you want to cruise over to see these rare paintings, American Watersports will take you. Otherwise, it's a tough six-mile hike from park headquarters.

At Lake Amistad, rules require use of designated boat ramps, and a small marina is convenient at each ramp. And as in many excellent national parks, you also have access to campsites, swimming areas (no lifeguards), and nature trails.

Lake Amistad's 850-mile shoreline, with its hidden canyons and stark landscapes, is great fun to explore. Those same prehistoric men who took shelter in Fate Bell have also been found in this area. Some of those holes in the steep canyons were their caves.

None of the shoreline has been developed for commercial use, so unless you camp, you'll have to stay at motels on the highway. Amistad Lodge, 512-775-8591, is the best of those close to the lake, or you can get away to desert solitude at Laguna Diablo Resort up on Devils River, 512-774-2422. At Laguna Diablo you have an apartment with *everything* furnished, even a washer/dryer.

The Del Rio area has a lot to offer for landlubbers, too. The old Val Verde Winery has been stomping grapes for over one hundred years, and just across the Rio Grande is Acuna, one of the best of the border towns. Don't miss a gourmet dinner at Lando's after browsing the shops. Then there's the ragged old Brinkley Mansion just down from the winery. "Doctor" Brinkley made a killing back in the thirties giving old men goat gland transplants to make them "frisky" again. At the Whitehead Museum you can visit the tomb of the famous old reprobate, Judge Roy Bean.

Lake Amistad is definitely the land of friendship, and you'll want to come back often, not only for the beautiful lake, but also for the famous Trans-Pecos hospitality.

CADDO LAKE

EAST

Marion County Chamber
of Commerce
West Austin Street
Jefferson 75627
214-665-2672

Amenities: Bed and breakfasts, boat tours, camping, cabin rentals, RV hookups

Actually, this mysterious East Texas lake should be named Kadohadacho, but Europeans contracted the name of the Kadohadacho Indians into Caddo. According to Caddo legend, a mighty chief was warned by the Great Spirit of impending disaster. The chief, heeding the vision, took his people to higher ground, whereupon the earth trembled, the ground sank, and floods poured over the land. But, fortunately, the tribe was saved. Unfortunately, they weren't saved very long. After the Texas Revolution, the white man wanted their land, and the Caddos were massacred.

In 1869 another tragedy occurred. The steamboat *Mittie Stevens* caught on fire on Caddo Lake. Had the sixty victims of the fire only known the water was only a few feet deep they could have waded to shore. What is left of the *Mittie Stevens* still rests in the thick mud of Caddo Lake.

Those were the days when Jefferson was a hustling, bustling river port, and big paddlewheelers came to her docks loaded with merchandise from all over the world. Little did Jefferson know that Big Cypress Creek was only navigable because of "The Great Raft." This was a massive stationary island of logs and every kind of river debris imaginable, estimated to extend 130 miles on the Red River. The jam was like a growing snake with its head catching every bit of debris that high water carried down from the two great forks of Big Red.

Captain Henry Miller Shreve was assigned the impossible task of destroying "The Great Raft," and he labored from 1833 until 1853 in vain. Finally in 1880 "The Great Raft" was broken, but it was too late. The train whistle had replaced the steamboat whistle. Jefferson, the town that defied Jay Gould and his demand for a railroad right-of-way, was literally left high and dry. Big Cypress Creek became a meandering stream, the trains roared in at Marshall, and Jefferson was stranded without access to the outside world. For many years the world forgot about Jefferson, but now it's the hottest little tourism town in Texas.

Folks just love to come and see the lovely antebellum homes, shop the antique stores, stay at the historic Excelsior House or one of the many bed and breakfast homes, and relive Jefferson's good old days. They can also relive the history of Caddo Lake and Big Cypress Bayou.

How about a canoe trip where steamboats used to churn the water? Nothing is more fun than slowly paddling your canoe down a clear creek as cypress trees heavily draped with Spanish moss block out the sun. Dense undergrowth and lush vegetation cover the shore, and silvery fish suddenly leap from the water. Among all this solitude it's hard to believe that Jefferson was one of the largest cities in Texas.

For this historic bayou trip, contact Mary Rose, 502 East Watson, Jefferson 75657, or call 214-665-7163. Her trips last one hour, half-day, or all day on the remote seven-mile upper section of the bayou. She also offers a three-hour tour of the Caddo Indian homeland and the Big Cypress Bayou steamboat route. Mary Rose knows just about all there is to know about the flora and fauna and history of this unique part of Texas. She's on the bank of the creek behind the McGarity Saloon at Dallas Street. Follow the dirt road to her bayou cabin.

Also available and extremely interesting are Chuck Nance's riverboat tours. The *Bayou Queen* even has moonlight serenade cruises. How's that for romantic? The *Bayou Belle* is for special charters with optional catfish dinners. Contact Captain Nance at Route 1, Box 1001, Jefferson 75657, or call 214-665-2222. Boats depart from Jefferson Landing and Depot at the corner of Polk and Bayou streets, and daily tours last about an hour and include all the good stories about the history of the bayou.

It's very hard to get a reservation at the famous old Excelsior Hotel, but you can probably get a gorgeous B&B room. Try the lovely Pride House or McKay House, and for a gourmet dinner, be sure to dine at Stillwaters.

For campers, the place to be is Caddo Lake State Park, Route 2, Box 15, Karnack 75661, or call 214-679-3351. This park is another one of the great legacies which the Civilian Conservation Corps built in the early 1930s. The cabins are wonderful, and the screened shelters are pleasant. Even though there are plenty of basic sites without utilities, remember those big East Texas mosquitoes.

Caddo Lake is ideal for fishermen and canoeists, but swimming is not recommended. Channels, or "boat roads," are endless, and wind around the lake offering a glimpse of almost prehistoric times. Hyacinths and water lilies add to the beauty of Caddo Lake. You can't help experiencing an eerie remoteness as you drift the boat roads, and even though the lake doesn't have a phantom "monster," you sense that something primeval lurks in its depths.

Over at Uncertain, Texas, it's business as usual with basic marinas and fish camps and restaurants serving East Texas fried catfish. Here, fishing is the order of the day, not history. Just get out there, cast your line, and catch one of those delicious bass—the bigger the better.

While Caddo Lake doesn't offer the traditional sailing and water-skiing, you should still put it on your list of water vacations. No other lake in Texas has so much murky beauty, history, and mystery. Stand by the water's edge and let the damp morning mist whirl about you, and you will be caught in the spell of Caddo Lake.

CRYSTAL LAKE

EAST

Box 488
Joaquin 75954
409-269-3259

Location: 11 miles east of Tenaha on U.S. 84

Admission Fee

Hours: 10:00 a.m.–7:00 p.m. Memorial Day through Labor Day

Amenities: Lifeguard, RV hookups, rest rooms, concessions

Restrictions: No alcohol

Talk about a Texas family tradition—do you ever find one at Crystal Lake! Orren Whiddon started digging out Crystal Lake in 1918 with oxen and slips, and he was still digging and damming up Spring Branch Creek well into 1919. His hard work was literally a big washout in 1933 when a flood swept the dam and Crystal Lake away. After World War II, Orren was looking for work and hit upon the idea of rebuilding the lake. With a bulldozer, his lake was ready for swimmers in three days. Modern machinery was definitely an improvement over oxen and slips.

Orren and his sons became slaves to their project, building a bathhouse, adding improvements, and modernizing and replacing facilities. So even though Crystal Lake opened for business in 1918, things really aren't the same—thank goodness.

When Orren Whiddon retired in 1980 and closed the park, even oldtimers shed some tears of nostalgia. In 1984 Orren's son, Abner, decided to follow in the family footsteps and reopened the park.

"Dad got the wild idea to put in a trolley. I think it was one of the first," Abner said. The innovative trolley was a smash hit! Squealers and screamers had the time of their lives riding that cable from a fifty-foot tower to plop ungracefully into deep water. So two more trolley lines were added, the tallest rising about seventy feet above the water. And the squealers and screamers are still having the time of their lives at Crystal Lake.

Tubes are available for the fainthearted, or you can swim out to big wooden rafts to soak up the rays. Lots of slides appeal to the younger set—the tough guys go down head first. The only flaw at this water paradise is that the older crowd will find it hard to ignore the disco music blaring at mega-decibel volume.

Next to Crystal Lake (the springs still pump out the cool flow) Abner has added a couple of modern water slides that are even more fun than the old slow trolley lines. This modern generation wants speed, chills, and thrills. you have to buy a separate ticket for the slides, and you can expect lines.

Legend has it that Bonnie and Clyde stopped by for a swim between murders, bank robbing, and kidnapping. They were probably worn out and wanted a cool break, but Orren says, "I wouldn't have known 'em if I'd seen 'em." Still, if Crystal Lake was on the Bonnie and Clyde tour, you know it's got to be good.

HIGHLAND LAKES

HILL COUNTRY

Highland Lakes Tourist Association
P.O. Box 1967
Austin 78767
512-478-9085

Amenities: Cabin rental, boat rental, concessions, camping, picnic area

It doesn't take long to get thoroughly and completely hooked on the Texas Hill Country. Maybe it's all the rocks, or maybe it's the scenic hills and winding roads that appeal to you. Or perhaps it's the clean dry air that is totally free of pollution. Probably it's all of these things, plus the stark beauty of the deep blue Highland Lakes.

The Colorado River rises in northeastern Dawson County, roughly between Lamesa and O'Donnell, and flows southeast about six hundred miles to empty into Matagorda Bay. Why it was named Colorado is unknown, as that's a Spanish word for red, and these waters are clear and blue. In 1685, Juan Dominguez de Mendoza christened the river San Clemente. La Salle named it La Sablonniere, but finally it became the Colorado. At one time in Texas history, the Colorado was a source of major transportation, but now its importance is as irrigation for the South Texas rice fields. Thanks to the rice farmers, Texas now has Lakes Buchanan, Inks, LBJ, Marble Falls, Travis, Austin, and Town Lake, all of which furnish some of the best recreation in Texas.

You could spend a lifetime on these lakes and never discover all the neat places to see, the good places to eat, the best fishing spots, the resorts and marinas, and the historic landmarks. There are so many attractions just a few miles from the lakes that you'll want to squeeze them into your visit, too. But no matter where you go, you can count on some of the friendliest people in Texas. To help you experience as much of this area as possible, *The Highlander* newspaper in Marble Falls has published a visitor's guide that comes out each spring and fall. You'll find the hunting hot spots (this is big deer country), the area's state and local parks, RV sites, and a complete guide to the history of the Highland Lakes. In addition, there's a directory of upcoming events, a dining guide, lodging guide, and information on churches. To receive a copy of this invaluable booklet, write *The Highlander* at P.O. Box P, Marble Falls 78654, or call 512-693-7367.

Have you ever wanted to take a boat trip to nowhere? Then contact the Vanishing Texas River people at P.O. Box 901, Burnet 78611, or call 512-756-6986. The best time to go is in winter when bald eagles soar above, shy deer are startled as they drink at the riverbank, and wild goats and pigs scamper back into cover. As the *Eagle II* chugs quietly up the completely deserted Colorado River, tourists are treated to a piece of unspoiled Texas. Cruises are daily throughout the year, but nature lovers prefer it between November and March when the bald eagles come to build their nests. In the summer there's a very romantic dinner cruise each Thursday, Friday, and Saturday night. How about a delicious steak dinner out in the middle of Lake Buchanan with a big moon overhead?

Another exciting choice offered by Vanishing Texas is a ride on the thrilling hydro-jet *Osprey* to Colorado Bend State Park for a picnic, hike, or swim. This guided tour lasts four hours, so bring your cooler and picnic lunch. Snacks and soft drinks are available on board. Don't forget your binoculars.

If you'd like to stay overnight, the owner, Ed Lowe, has opened two bedrooms in his lovely home for bed and breakfast, so you can stay right on Lake Buchanan (locals pronounce that BUCK-han-an).

Just down the road on Lake Buchanan is the Knittel House, P.O. Box 261, Buchanan Dam 78609, 512-793-6409. This modern home is right on the banks of the lake, and if you bring your own boat you can use the Knittel House dock. Accommodations include a huge bedroom with private bath and a full breakfast. There's also an easy-living patio and swimming pool which makes Knittel House a perfect retreat spot.

The really big-time resort on Lake LBJ is Horseshoe Bay—and it is gorgeous. Talk about manincured landscaping! Every blade of grass is the perfect height, and every bush is meticulously groomed at Horseshoe Bay. Out by the tennis courts is an absolutely fabulous Japanese garden with lovely fountains and waterfalls. It is almost a hidden treasure, and many guests miss it entirely, as it is tucked behind the tennis pavilion.

Horseshoe Bay is the place where the wealthy people go. Everything is elegant—the small hotel with continental cuisine, the dining rooms (always a dress code), and two swimming pools (one for adults only). Not one, not two, but three golf courses are part of this resort! The Slick Rock, Ram Rock, and Apple Rock courses were all designed by Robert Trent Jones, and each has been perfectly landscaped. All players must be club members, guests of the resort, or guests of club members, and they may take advantage of any of the three challenging courses.

The marina is fully equipped to handle any aquatic needs: boat and ski rentals, boat sales and repair, and, of course, fishing guides. Don't bother to drag your boat along—just buy one when you get there.

Horseshoe Bay offers all sorts of vacation packages, and if it's worth doing on a vacation you can do it here. You can fly into Horseshoe Bay in your private plane, and a red carpet will be rolled out to take you to the resort. Describing this resort demands all the superlatives. Horseshoe Bay could be your big splurge vacation or second honeymoon.

Contact: Horseshoe Bay Country Club Resort, P.O. Box 7766, Horseshoe Bay 78654, or call 800-252-9363 Texas, 800-531-5105 nationwide.

If you want to be close to the big city of Austin, Lake Travis is your best choice. And speaking of choices, there are unlimited choices for your water vacation on this popular lake. The Lakeway Inn claims to be "The Most Famous Resort in Texas." And there's no doubt that it lives up to its claim. How about a full-service marina with everything that floats available for rent, including party boats? You can sail, fish, ski, paddle, or just party from this marina. If that gorgeous lake water is a bit chilly, try one of the Inn's three swimming pools. Then reserve an early morning tee time at one of the two championship golf courses.

There are thirty-two tennis courts on which to exhibit your great backhand. If tennis isn't your game, trot over to the stables and sign up for a canter on twenty-five miles of heavily wooded Hill Country trails.

As for romance, the Inn's rooms have balconies overlooking the lake, and the world looks beautiful indeed when the moon glimmers over that smooth silver water. In the cool months, ask for a room with a fireplace.

The dining room has a dress code, and before 7 p.m. there is a special lower rate for their excellent dinners. Lots of small conventions like Lakeway Inn's facilities, too, so reservations are a must. You'll find the rates very reasonable for this first-class hotel, even in the height of summer. Contact The Lakeway Inn at 101 Lakeway Drive, Austin 78734, or call 1-800-LAKEWAY.

If you prefer to rent a whole house at Lakeway, you can bet it will be very nice indeed, because people who build their homes here have a high standard of living. Rates are lower in the winter and pets are not allowed. For example, a three bedroom, two-bath home close to the marina, with a jacuzzi in the courtyard, is $230 a day in the summer and $200 a day in winter.

But maybe you never want to leave the water at all. If so, the perfect solution is to rent a houseboat. A forty-eight-foot-long houseboat that sleeps ten with *everything* furnished is very reasonable. During the week is less expensive than weekend rentals. You'll find the perfect houseboat and every other kind of boat for rent at Hurst Harbor Charter on Lake Travis at 16405 Marina Point, Austin 78734, 512-266-1069.

Another very nice resort on Lake Travis is The Inn on Lake Travis, which has a swimming pool and boat launch. The restaurant serves mainly steaks and seafood, and is open for breakfast and lunch. Golf courses are nearby. Call 800-252-3040.

Somewhat like Lakeway in its amenities, but on the opposite side of Lake Travis, are The Resort and Country Clubs of Lago Vista. They have golf, tennis, swimming, boating—just about everything except horseback riding.

You can chose a hotel room, condominium, or houseboat to knit up those jangled nerves. Austin, with its historic buildings and fine restaurants, is about thirty minutes away if you yearn for some sightseeing. Contact Lago Vista Country Clubs, P.O. Box 4871, Lago Vista 78645, 512-267-1131.

Highland Lakes has every type of accommodation you can possibly want—in every price range. Your best bet is to contact the Highland Lakes Tourist Association or get a copy of *The Highlander*'s Visitor's Guide for complete information. Also request a copy of their Bluebonnet Trails. In the spring, the Texas Hill Country is a sea of blue—even bluer than its lakes—when bluebonnets cover the hills and create an unforgettable picture.

LAKE LIVINGSTON

SOUTHEAST

Waterwood National Resort
and Country Club
Waterwood Box 1
Huntsville 77340
409-891-5211

Amenities: Camping, picnic area, RV hookup, cabin rentals, rest rooms, showers

When the Trinity River was dammed, it created one of the largest lakes in Texas. Lake Livingston is in a beautiful setting with tall, dark green pines surrounding its miles and miles of shoreline. It had all the ingredients for becoming the Lake Travis of East Texas, but somehow this never happened. A few nice subdivisions were built on the shores and a few adequate motels opened, but Lake Livingston became the proverbial fisherman's paradise, and has remained that way. Marinas and lodges are the principal accommodations, and there is little for the nonfisherman to do. For all the information, contact the Onalaska Chamber of Commerce at 409-646-5000.

For campers, there is the excellent Lake Livingston State Recreation Area with every amenity a camper could possibly want. Go for one of the screened shelters; the mosquitoes get pretty big in East Texas. If dipping the pole isn't for you, the area has an excellent pool, lots of hiking trails, smooth water for balancing on water skis, and a playground for the little guys.

Fortunately, the lake also has one perfectly beautiful resort—Waterwood. Halfway between Livingston and Huntsville on U.S. 190 is the striking entrance to the Waterwood world of extraordinary natural beauty and tranquility. Don't drive too fast as you enter, because families of roadrunners make this drive their home, too.

You know you are heading for a special place on the long drive into the deep woods, and it's rather a shock to see small street signs, but people do live on this side of Lake Livingston. Well-marked roads lead to the marina, condominiums, and golf course that blend so perfectly into the landscape.

The Cabanas and Lodge offer eighty-four beautifully appointed guest rooms with patios and balconies overlooking the golf course. Or you may prefer rooms on Pools Creek Park, secluded hideaways with all the luxury of any of the guest rooms. All have very reasonable rates.

If golf is your game, the eighteen-hole USGA championship course is one of the top six courses in Texas. Every golfer should tee up and test his skill here. You have four lighted tennis courts. Or how about a dip in one of four pools? Don't forget your workout at the Waterwood Health Club, which has hydraulic weight equipment, saunas, and aerobic classes.

You'll find the elegant Garden Room a gorgeous setting for leisurely meals of fine cuisine with attentive service. For cocktails, there's the adjourning Garden Court Lounge. Both have wonderful views of the golf course and surrounding forests. Fat ducks and geese waddle complacently across the grounds, and you amateur bird watchers can just sit on your balcony and do some sightings.

The marina is on a quiet cove near the RV park. Waterwood residents and guests find the docks and piers an excellent spot for catching a big one for dinner. The marina is pretty

Lakes Superior

basic, having no restaurant and no boats for rent. If you plan on sailing or water-skiing, you must bring your own boat. They do offer wet storage, so you can leave your boat in the water all weekend.

Lake Livingston is so big that you should pick up a map at one of the marinas or convenience stores on the highway, so you can really enjoy exploring its shores and coves. A great boat trip is to head up the lake to the Trinity River. Somewhere in the Trinity River Authority's files is a master plan for the river to be made navigable all the way to Dallas. But the restaurants and service stations on I-45 aren't worried about being replaced by the marinas in the foreseeable future.

LAKE TEXOMA

NORTHEAST

Sherman Convention
and Visitors Bureau
306 N. Travis
Sherman 75090
214-893-1184

OR

Chamber of Commerce
313 W. Woodard
Denison 75020
214-465-1551

Amenities: Boat rentals, camping, RV hookups, picnic area

When the Red River was dammed in 1944, a 100,000-acre lake was created on the borders of Texas and Oklahoma. What better name for this new man-made lake than Texoma? It has the distinction of being the only works project completed during WWII. All other manpower and materials went to the war effort. Like any lake of this vast size, marinas, boat ramps, and motels abound along its miles of shoreline. And just think of all those other miles of open water for cruising! And what about all the miles of undeveloped shoreline with countless beaches and coves to explore? So, no matter what's your pleasure, on Lake Texoma there's plenty of room to do it.

To really get the most out of your Texoma trip, why not call Willow Springs Resort and Marina and say, "Hey, reserve me a houseboat, a big one. I've got fourteen people that want to go. Can you do it?" Well, they certainly can, if you call far enough in advance to 405-924-6240. They also have houseboats that sleep ten. No matter how many go, you're

Lakes Superior

all going to have a wonderful time on one of the biggest lakes in Texas.

You can fish and swim from the houseboat, tow your fast ski boat or fishing rig, and pick a new destination each day. If you don't own a fast ski boat or fishing rig, Willow Springs is well-equipped to rent you one of those, too. You are on your own, and you set the pace and schedule *you* want. Explore and wander. Your only earthshaking decisions are when to eat, where to stop, and whether you should fish, swim, or ski. You don't even have to decide what to wear. Just put on your bathing suit and forget it. Who needs a swanky wardrobe on a houseboat?

At Willow Springs you are checked out on the operation of the boat—start-up, maneuvering, docking, mooring, marine radio, depth finder, lights, and safety gear. You are definitely the captain of your ship—well, your houseboat. To moor your vessel, just locate a pretty beach, idle up to the shore, nose the steel hull onto the sand, and tie that boat to a couple of trees. Easy? You can bet on it.

Your vacation liner has everything you left behind at home except bedding, towels, and groceries. If you are a real softie, bring along a stereo, some cassette tapes, and a TV. Oh, and don't forget those nine or thirteen good friends to share all the luxury that includes a full bath, a half-bath, and air conditioning.

As for dining at the captain's table and having an end-of-cruise masquerade ball, you'll have to work that out on your own. Rates vary depending on the boat size and the time of year. A deposit and charterers' insurance are required, and there is a stiff cancellation-of-reservation penalty.

Willow Springs believes that "houseboating means . . . FREEDOM!" and you'll probably be inclined to agree with them after a vacation on one of their luxury liners. Contact Joe and JoAnne Showalter at Willow Springs Resort and Marina, Rt. 1, Mead, OK 73449 or 405-924-6240. They are just south of US Highway 70, ten miles west of Durant, Oklahoma. Bon Voyage!

```
R0162977019 txr          T
                         333
                         .784
                         R922
Ruff, Ann, 1930-
Texas water recreation /
     Ann Ruff
```